Pesticides on Plant Surfaces

Critical Reports on Applied Chemistry Volume 18

Pesticides on Plant Surfaces

edited by Helen J. Cottrell

Published on behalf of the Society of Chemical Industry by
John Wiley & Sons
Chichester · New York · Brisbane · Toronto · Singapore

British Library Cataloguing in Publication Data:

Pesticides on plant surfaces.——(Critical
  reports on applied chemistry; 18)
  1. Pesticides   2. Plant surfaces
  I. Cottrell, H.J.   II. Series
  632'.95   SB951

ISBN 0 471 91478 9

Library of Congress Cataloging-in-Publication Data:

Pesticides on plant surfaces
  (Critical reports on applied chemistry; v. 18)
  Includes index.
  1. Plants, Effect of herbicides on.   2. Plants,
Effect of insecticides on.   3. Plant surfaces.
4. Herbicides.   5. Insecticides.   I. Cottrell, H. J.
II. Society of Chemical Industry (Great Britain)
III. Series.
QK753.H45P47   1987   632'.95   86-32536

ISBN 0 471 91478 9

Typeset by Activity Ltd., Salisbury, Wiltshire
Printed and bound in Great Britain

Contents

List of Contributors

M.G. Ford   Department of Biological Sciences, King Henry Building, Portsmouth Polytechnic, King Henry I Street, Portsmouth PO1 2DY

R.C. Kirkwood   Biology Division, Department of Bioscience and Biotechnology, University of Strathclyde, Glasgow G1 1XW

D.W. Salt   Department of Mathematics and Statistics, Hampshire Terrace, Portsmouth Polytechnic, Portsmouth PO1 2EG

Editor's introduction

It is commonplace that the outsides of things are different from the insides, as we prove when we cut off the cheese rind or peel an orange before eating them. Chemists have long known that certain molecules behave differently at the surface of a solution from those in the bulk. Biologists, on the other hand, have always understood that the surface of an organism is its first line of protection against external forces and alien bodies. The unusual properties of surfaces provide living organisms with a protective layer which modifies the exchange of chemicals, including pesticides.

The applied science of crop protection, which is barely half a century old, studies among other problems those arising from the application of biologically active chemicals on to the surfaces of plants, with the objective of removing unwanted species or of controlling insect and fungal attack. This has created interesting interfaces between a number of quite different scientific disciplines. It was with the object of encouraging the study of such interfaces that the Physicochemical and Biophysical Panel of the Pesticides Group of the Society of Chemical Industry was originally instituted about twenty years ago by some far-seeing chemists, led by Dr G.S. Hartley. They believed that the exchange of ideas and research findings between scientists of different specializations would stimulate further investigation into basic problems of the physical and chemical aspects of biological activity. Progress in this area has justified their optimism and the journal *Pesticide Science* was started in 1970 to publish relevant papers. This review volume is a tribute to these pioneers, although it covers only a fraction of their original interests. Steady progress in surface research has been made during the last two decades, with quite spectacular results obtained during the last ten years. This has been achieved by the combined efforts of chemists, biochemists, physicists, botanists, entomologists, computer scientists and mathematicians, and has involved as well as more traditional methods, the use of many modern research tools, especially spectroscopy, electron microscopy, radio chemical synthesis, autoradiography, liquid scintillation counting, chromatography and insect behavioural studies.

As often happens, the discovery of a new generation of agrochemical products was telescoped into the few years corresponding with the Second World War. Thus DDT, 2,4–D, γ-BHC and IAA greeted the peace of 1945 and were enthusiastically embraced by practical farmers and experimental scientists alike. Both privately and publicly funded research organizations joined in the search for new and better insecticides, herbicides, growth regulators and fungicides, and these were forthcoming in plenty. The commercial rewards of marketing a successful agrochemical were not inconsiderable, given the drive by national governments and international agencies to maximize agricultural production, and given also the introduction of economic measures to stabilize the prices of agricultural commodities.

The consequence of all these factors was to increase the money available for agrochemical research, and some resources went to study how the most efficacious products worked. Apart from the element of just wanting to know why, such research programmes could be and were justified on the grounds that, if the mechanisms of action were better understood, this would help to design even more effective products and to use existing ones more effectively. Early experimenters tended to follow classical biochemistry and used isolated cells to study the effects of pesticides on plant processes, thereby bypassing the protective action of outer cells, leaf cuticles and even cell membranes. However, there were some honourable exceptions to this trend who studied the way in which the new insecticides penetrated the cuticle of insects and others who began to study the structure and composition of the plant cuticle, its waxes and the characteristics of pesticide retention and penetration.

The contents of this review volume are concerned with the behaviour of two types of products, herbicides and insecticides, which are normally used in practice as aqueous sprays directed on to the leaves of plants. It was originally hoped to include a review of the behaviour of insecticides either on inanimate surfaces or on animal skin, hair and fleece, but this proved not to be feasible because of the paucity of published work on the general topic, and more particularly on the newer insecticides that appeared to be of interest for public health and animal health applications. It is difficult to escape the conclusion that such work is still awaiting publication or perhaps has only appeared in rather obscure journals. I mention this fact in the hope that some influential persons may read this and take the necessary action to bring the results of such studies into the public sphere.

In common with previous Critical Reports, the authors have been more concerned with assessing the quality of the work reviewed, a task for which all of them are well suited since all are actively engaged in relevant fields of research, than with compiling comprehensive bibliographies. The present review is arranged in two chapters. The first, by R.C. Kirkwood, discusses the uptake and movement of herbicides from plant surfaces and the effects of formulation and environment upon these processes, while the second, jointly by M.G. Ford and D.W. Salt, describes the behaviour of insecticide deposits and their transfer from plant to insect surfaces.

Chapter 1 begins with a very full review of the present state of knowledge about the chemical composition and fine structure of the outer layers of leaf surfaces and a description of the factors that have been found to affect the penetration of herbicides into the underlying cells. Unfortunately the section on the effects of formulation on cuticle permeability is less ample, although recent papers suggest that there is increased interest in such studies. Similarly, the section on the effects of environmental factors reflects the comparative difficulty of carrying out satisfactory experiments, since the author stresses the importance of pretreatment growth conditions on experimental plants. However, the discussion about sites of preferential entry of herbicides is dealt with as fully as possible, although it is a controversial issue that remains unresolved. The section on the absorption of herbicides by plant tissues is reviewed by reference to kinetic studies and a discussion of possible mechanistic models. The author also considers rival classical theories which seek to explain the movement of chemicals in the conducting tissues of plants, since it is clear that translocation from the region of absorption must have an ultimate effect on the kinetics of cuticle penetration, tissue absorption and movement. Similarly, he has also briefly reviewed evidence on the effects of plant metabolic processes upon herbicide molecules because of their importance in the interpretation of experimental data.

One conclusion stands out very clearly on reading this chapter: the danger inherent in generalizing the results from one plant species to another and from one herbicide to another. In this context, the limited applicability of general theory is a signal of hope for those who aim to discover novel herbicides with more selectivity of action than known ones. A further interesting point that emerges is that only a very small area of the entire plant needs to be covered by spray deposits of a translocated herbicide to obtain optimum biological efficacy. As will be seen, this contrasts sharply with requirements for insecticide spray coverage.

The second chapter on the behaviour of insecticides introduces a third component into the system — the insect or acarine pest. The authors begin by describing briefly the methods of formulation, application and spray impaction studies on both plant surfaces and the insects themselves. Since the study of cuticle structure is common to both fields of work, Ford and Salt discuss only the effects of topography and superficial structure on the deposition of sprays, and refer readers interested in chemical composition to the appropriate section in the first chapter. It is evident from the start that the requirements for insecticidal efficacy are different from those for herbicidal efficacy, since it is extremely important that insecticides remain available to crawling or biting pests for as long as possible by staying on the outside of the leaves. Hence the review of work on losses of insecticide by physical and chemical processes is vital. The section on the transfer of insecticide deposits from plant surfaces on to insects covers a relatively long period of work and includes some early but classic studies with the chlorinated hydrocarbons as well as the latest findings, and the authors' suggestion of a mathematical model for the transfer of insecticides. For completeness the authors also review studies on the transfer of insecticides by direct impaction on to insects, by ingestion and by toxic vapour. The remaining sections of the chapter review very recent and imaginative research on the influence of insect behaviour on pickup and on the relationship of spray application characteristics to biological response. It is particularly encouraging to see that some workers are attempting to simulate the field situation under laboratory conditions, so as to study the pickup of toxicants by both sessile and mobile insects. Such field experiments are notoriously difficult to perform successfully, and the review of attempts to predict efficacy from the distribution of spray deposits by means of a computer model is very interesting. The authors also review the results of field studies on the performance of spray deposits from electostatically charged atomizers.

I hope that in this introduction I have drawn attention to the similarities and differences in subject matter as well as differences which characterize each chapter. Thus, Kirkwood is interested in the transfer mechanisms of herbicides from plant surfaces to the internal sites of action, while Ford and Salt remind us that the insecticide must remain on the surface to be available to the pests moving on the leaf or eating it. Although no quantitative estimates are given, the implications are that insecticidal deposits must be as nearly continuous over the surface as possible, although in some circumstances it has been said of an insecticide that 'if it doesn't work then use less'.

Much of the work reviewed in this book is of a preliminary or incomplete nature and some of the findings may be challenged by future work. However, I believe it is true to say that such studies are extremely valuable both to applied and pure scientists. The pesticide scientist will take heart from the knowledge that it is possible to improve the deposition of sprayed materials on plant surfaces and also to change the physical form of spray deposits by practical measures such as the design of spray machinery and the chemistry and physics of formulation. The ability to change at will the behaviour of biologically active molecules to suit them to the required task and to make their performance more predictable under natural conditions are still pipe dreams, yet we are conscious of making progress in those directions.

On the other hand, the academic scientist should reflect that discoveries about the fine structure and behaviour of living surfaces would have been much slower or might not yet have occurred had it not been for the impetus of a thriving agrochemical industry. Once again the needs of applied science appear to have stimulated fundamental discoveries about the properties of matter.

This book is intended for all practitioners of the science of crop protection and teachers and the more advanced students of crop protection as a useful reference book and guide to the more fundamental study of how biologically active compounds interact with the living surfaces of plants and insects. It includes well-researched topics, more controversial theories and, even, some untidy ends and neglected areas of enquiry.

Helen J. Cottrell

Glossary

Editor's Note: Pesticides are generally referred to in the text by their approved common names, as listed in the *Pesticide Manual* published by the British Crop Protection Council. However, the glossary below gives the chemical names of those commonly referred to only by a group of characters, such as 2,4-D and DDT.

Abaxial — the leaf surface furthest from the axis of plant growth
Acer pseudoplatanus — sycamore (tree)
Adaxial — the leaf surface nearest the axis of plant growth
Agrostis tenuis — common bent grass
Allium porrum — leek
Angiosperms — flowering plants
Apoplast — the non-living matter of cells
Arthropods — invertebrate animals with jointed legs and a hard external skeleton. Includes insects and spiders
ATP — adenosine triphosphate
Auxin — indol-3-ylacetic acid, a naturally occurring plant growth regulator
Avermectins — a family of products from *Streptomyces avermitilis* with activity against arthropods

γ-BHC — *see* γ-HCH

CDA — controlled droplet application; a spray technique which produces a narrower range of droplet sizes than conventional hydraulic machines
Chloroplasts — discrete bodies in plant cells which contain chlorophyll
Chrysosporium pannorum — free-living fungus of soil, plant litter and leaves
Citrus auranticum — Seville orange
Cytoplasm — the living part of a cell

2,4-D — (2,4-dichlorophenoxy)acetic acid
2,4-DB — 4-(2,4-dichlorophenoxy)butyric acid
DDT — 1,1,1-trichloro-2,2-bis(4-chlorophenyl)ethane
Drosophila melanogaster — fruit fly

Ectodesmata — strands of protoplasm linking cells
Electrostatic application — a technique by which spray droplets are given an electrostatic charge

Elytra — a pair of scales enclosing the flying wings of insects

EO — oxyethylene units (-CH$_2$CH$_2$O-) appearing in many surfactants, e.g. Tween

EPTC — S-ethyl dipropylthiocarbamate

Gramineae — family of monocotyledons including cereals and grasses

γ-HCH — 1,2,3,4,5,6-hexachlorocyclohexane; activity is mainly due to the γ-isomer, gamma-HCH (synonym gamma-BHC)

IAA — indol-3-ylacetic acid

Instar — a stage in insect metamorphosis

Larva — the immature stage of an insect, e.g. a caterpillar

Lemna minor — duckweed

Lepidoptera — moths, butterflies

Lissapol NX — non-ionic surfactant

Malus hupehensis — ornamental crab apple

Malus pumila — cultivated apple

MCPA — (4-chloro-2-methylphenoxy)acetic acid

MCPB — 4-(4-chloro-2-methylphenoxy)butyric acid

Mesophyll — layer of chloroplast-containing cells between the epidermal layers of a leaf

Microencapsulation — a technique in which the active ingredient of a formulation, with or without adjuvants, is enclosed in minute capsules which remain suspended in an aqueous medium, enabling it to be diluted like a conventional concentrate and applied with a spraying machine

Mitochondria — minute bodies in the cytoplasm of a cell which are responsible for respiration and specialized reactions

Myzus persicae — the peach potato aphid

NAA — 1-naphthylacetic acid

NP — nonylphenol, used in surfactants, e.g. 'NPE' — nonylphenol polyethoxylate

OP — octylphenol, used in surfactants

Organelles — small bodies within cell cytoplasm, having specialized functions, e.g. chloroplasts, mitochondria

Palisade cells — a layer of elongated cells, rich in chloroplasts, lying below the epidermis and perpendicular to the plane of a leaf

Phloem — the living conducting tissue of a plant by means of which assimilates travel from the leaves to other parts

Photosynthesis — conversion of CO_2 and H_2O to sugars, etc., by chlorophyll and solar energy

Phytophagous — plant-eating (pests)

Pieris brassicae — cabbage white butterfly/caterpillar

Plasmalemma — the outer cell membrane

Plodia interpunctella — a lepidopterous pest of fruits and stored grains

s.m.c. — soil moisture content

Spans — non-ionic surfactants, fatty acid esters of sorbitan

Spodoptera littoralis — cotton leafworm; an important pest of cotton and other crops

Stomata — pores in the leaf epidermis through which gaseous exchange occurs. They are bounded by specially adapted guard and accessory cells

SVC — saturation vapour concentration

Symplast — the living matter of cells; the cytoplasm

2,4,5-T — (2,4,5-trichlorophenoxy)acetic acid

Tarsal setae — bristles on the legs and feet of some insects, especially flies

Tergitol — non-ionic surfactant

Trichome — epidermal outgrowth, e.g. leaf hair

Trifolium repens — clover

Triton X — a non-ionic surfactant; polyethoxylated octylphenol

Tweens — non-ionic surfactants chemically resembling Spans, but rendered more hydrophilic by the addition of a polyoxyethylene chain

$U^{14}C$ — lebelled unspecifically with ^{14}C

ULV — ultra-low volume (of spray applied)

Vacuole — cell cavity containing air, liquids, food, waste products, etc.

Xylem — non-living conducting system of plants by which water and minerals are conducted from the roots to leaves, etc.

1 Uptake and movement of herbicides from plant surfaces and the effects of formulation and environment upon them

Ralph C. Kirkwood
University of Strathclyde, Glasgow

1.1 Introduction

The activity of a foliage-applied herbicide must ultimately depend on the concentration of active ingredient that reaches the sites of action together with the effect of the herbicide on the biochemical mechanisms taking place at these sites. A number of interrelated factors may influence its uptake and movement, including the efficiency of cuticle retention and penetration, tissue absorption

1

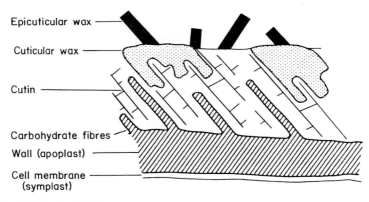

Epicuticular wax

Cuticular wax

Cutin

Carbohydrate fibres

Wall (apoplast)

Cell membrane
(symplast)

Figure 1.1. Idealized cuticle[5].

and, in the case of systemic compounds, phloem 'loading' and translocation; herbicide metabolism or immobilization en route may reduce the amount reaching the active sites. Interspecific differences in the efficiency of these factors may have important effects determining herbicide selectivity. In this review it is intended to consider the importance and role of factors influencing the uptake and translocation of foliage-applied herbicides, with particular reference to the influence of formulation and environmental factors.

1.2 The cuticle

The surface of the aerial parts of plants is covered by a lipoidal cuticle which minimizes water loss from the plant and also may act as a barrier to the penetration of foliage-applied chemicals, particularly those of a polar nature. The structure of the cuticular membrane has been reviewed recently by Holloway[1]. The outer wall of the epidermal cells is characterized by the presence of a cuticular membrane composed of a framework of polymeric cutins with embedded cuticular wax and soluble waxes deposited on the surface as epicuticular wax. The cuticular membrane is not structurally or chemically homogeneous but is composed of a number of layers, each of which is defined in respect of its position and chemical composition. The characteristics of these layers vary according to species and stage of development and this should be borne in mind when considering a generalized scheme for the structure of the plant cuticle membrane (Figure 1.1). Carbohydrate fibres may extend from the underlying wall and middle lamella into the cuticle[2, 3], and where these extend from the aqueous apoplast almost to the external surface of the cuticle[4] they may provide an aqueous continuum traversing the cuticle[5].

The chemistry of the plant cutins has been reviewed by Holloway[6]. They are insoluble high molecular weight lipid polyesters composed of long chain substituted aliphatic acids combined with hydroxyl (mainly), aldehyde, ketone, epoxide and unsaturated groups. In Angiosperms the cutins can be divided into three types according to chain length of the dominant monomers: C_{16}, C_{18} or a mixed C_{16} and C_{18} type. The most common C_{16} compounds are 9,16- and 10,16-dihydroxyhexadecanoic acids, while the most abundant C_{18} compounds are 9,10-epoxy-18-hydroxy and 9,10,18-trihydroxyoctadecanoic acids. The composition of the cutin monomers varies according to plant species and position on the plant, probably being derived mainly from the biotransformation of hexadecanoic and octadecanoic acids[6].

The physicochemical properties of the epicuticular waxes are of paramount importance in determining wettability of the leaf surface. The crystalline structures of which the waxes are composed are superimposed on the surface of the cuticular membrane and may take the form of plates, tubes, ribbons, rods, filaments or dendrites[7]. The configurations of epicuticular wax deposits vary according to species, surface and age of leaf. Thin films, thick amorphous layers and plate waxes are frequently found whereas tubes and other crystalline structures are restricted to a small range of species. Generally, crystalline wax is arranged uniformly over the leaf surface, though this may constitute only a small proportion of the surface deposit[8]. The morphology of these structures is determined mainly by the composition of the wax exudates carried through the cuticular membrane in a volatile solvent and not by a mechanism of wax extrusion through micropores[9–11]. Wax structure is modified by changes in solution concentration, and environmental effects may influence crystallization rates and thus modify wax structure.

The epicuticular waxes are composed of a variety of long chain aliphatics, pentacyclic triterpenoids, sterols and flavonoids; relatively few of these commonly form major components, examples being primary alcohols (C_{26}; C_{28}; C_{30}), hydrocarbons (C_{29}; C_{31}), secondary alcohols (C_{29}) and β-diketones (C_{31}; C_{33})[7]. In contrast, the cuticular (embedded) waxes are composed principally of fatty acids of the 'short series' (C_{16}; C_{18}).

1.3 Surface retention and physical losses

The amount of herbicide deposited on a crop/weed canopy at the time of application is influenced by morphological, spray, formulation and environmental factors. The 'plant' factors include the habit of growth, the size, shape and orientation of the leaves, and surface characteristics such as corrugation, the presence of hairs and the physicochemical properties of the epicuticular waxes. The importance of the epicuticular wax as a barrier to cuticular penetration of many substances is well established[12–14]. Spray deposition is generally proportional to the rate of application (in kilograms per hectare) with losses occurring due to runoff, volatility or spray misdirection; the height of the boom above the canopy and the location of the leaves in the canopy may also influence surface deposits. Spray volume and droplet size are also of considerable importance, presumably due to interaction involving such factors as retention, spreading, wetting and penetration of the surface of the foliage[15]. In general, herbicidal efficiency increases with reduction in droplet size[16, 17], and optimal effects may be obtained when maximum contact area is obtained by combination of small droplets and high volume[18]. However, large droplets may be preferable in situations in which fine hairs are present on the plant surface[19] or where there is a need to exploit differential retention between a waxy crop and less waxy weeds or between broad-leaved and narrow-leaved species[20].

The optimum volume and spray droplet size may be influenced by the phytotoxicity of the herbicide (and vice versa). Generally, translocated compounds can be sprayed at relatively low rates per hectare (ha) without loss of activity. For example, comparison of droplet sizes (150–350 µm) and spray volumes (5–45 l/ha) has shown that translocated herbicides can be sprayed at rates as low as 20 l/ha without loss of efficacy[21]. Similarly, no major loss of effectiveness was obtained when 2,4-D ester or Barban were applied at 5–20 l/ha compared with 165–200 l/ha[22].

Controlled droplet application (CDA) has been effective in the case of systemic compounds, but less so in the case of contact herbicides. For example, in greenhouse trials, at a droplet size of 250 µm, lower volumes of CDA sprays (5–45 l/ha cf. 200 l/ha) have been effective in the case of MCPA, mecoprop, dicamba, 2,3,6-trichlorobenzoic acid, Barban, difenzoquat and chlorofenprop-methyl; the contact herbicides ioxynil, bromoxynil and bentazone performed less well, perhaps because of inadequate covering[23]. The increased concentration of active ingredient (a.i.) per droplet may have affected uptake of the herbicide due to phytotoxic effects. When spraying at a volume of less than 25 l/ha a coverage of at least 1 per cent for translocated herbicides and 5 per cent for contact herbicides has been recommended[24].

Penetration of droplets into dense crop canopies has caused problems due to the relatively small number of droplets which penetrate[15]. This may be partly overcome, however, by spraying into an airstream created from the ambient air flow diverted by an aerofoil into the crop[25].

Formulation can also be important in spray deposition. For example, lower deposits are often obtained with wettable powders compared to emulsifiable concentrates (E.C.) or flowable formulations[26]. The use of E.C. formulations or the inclusion of oils or adjuvants appear to minimize the early loss of a pesticide by increasing the rate of penetration. Spray deposits and thus the amount of chemical available for uptake are influenced by environmental factors such as wind, rain, light and temperature and also by interaction with the complex microenvironment of trichomes, secretory glands, stomata, epicuticular wax and cuticles[27].

The factors which affect dissipation of a pesticide after application have been reviewed by Ripley and Edgington[26]. They have developed a modelling approach to explain the decrease in surface residues caused by volatilization, hydrolysis on photodecomposition or to physical properties such as vapour pressure, water solubility or partition coefficient. Disappearance curves for most foliage-applied pesticides show an exponential decline, and first-order reaction kinetics have been used to explain the decline, though several different or mixed-order rate equations may operate over the lifetime of the pesticide. Account should be taken of the factors that may cause losses in the initial period after deposition, such as volatility or runoff and the possibly long persistence of low concentrations in the plant. While it may be possible to obtain good correlations for time or weather variables alone, it may be impossible to ascertain the relative importance or contribution of a particular variable. In a correlation analysis, anomalous data points should be considered in terms of physical or environmental processes such as volatilization, vapour pressure or washoff redistribution[26].

1.4 Cuticle penetration by herbicides

The physicochemical properties of the cuticle in relation to herbicide absorption have been reviewed by Hull[28], Martin and Juniper[29], Kirkwood[30] and Price[5]. Cuticular permeability is influenced more by the wax component than the cutin matrix[14], and the importance of the lipophilic route is implied by the strong correlation which exists between the lipophilic nature of the herbicide and its ability to penetrate the cuticle[31, 32]. This relationship has been observed both for different herbicides and different formulations of the same herbicide. It is also reflected in the faster uptake of the phenoxyacetic acids at low pH values when the molecules are in the non-polar, unionized form[32–34]. It seems reasonable to conclude that the lipophilic route across the cuticle is quantitatively much more important than the hydrophilic pathway, at least in the case of the phenoxyacetic acids[35].

The mechanism of cuticle penetration and the influence of various factors have been reviewed by Price[5]. The process is believed to involve diffusion, the rate being described by Fick's first law modified to membranes[36]. In considering the factors influencing the penetration of pesticide through the leaf cuticle, Price[5] substituted the Stokes–Einstein equation in Fick's first law as follows:

$$J = \frac{k^\beta \, T \, K}{6\pi r \, \eta \, \Delta x l} \, (C_o - C_i)$$

where
J = flux
k^β = Boltzmann constant
T = absolute temperature
K = partition coefficient
r = radius of the solute molecule
η = viscosity of the solvent
Δx = thickness of the cuticle

l = tortuosity factor
C_o = concentration of solute outside the cuticle
C_i = concentration of solute in the wall inside the cuticle

1.4.1 Cuticle thickness

Cuticle thickness has been reported to be a major factor influencing cuticle penetration; thus thick cuticles would be expected to be less permeable than thinner ones. For example, Scheiferstein and Loomis[37] and Hull[38] reported that the rate of penetration was correlated inversely with cuticle thickness. This is in accord with Fick's second law which states that the time for diffusion increases with the square of the distance[36]. However, the results of Norris[39] and Bukovac et al.[40] suggested that the composition and arrangement of the cuticular components was of greater importance than thickness in determining permeability. The latter found changes in cuticular permeability during expansion of peach leaves and attributed these partly to the influences of variations in the composition and distribution of the surface wax.

The effect of age on the development of epicuticular wax layers has been studied by several workers. The thickness and composition of epicuticular wax varies with leaf age[41–44], the amount of wax per unit area generally fluctuating during growth with an overall decline occurring on most leaves. Baker and Hunt[8], investigating development changes in leaf epicuticular waxes in eight plant species, found that changes in wax production occurred during a critical 3-week period following leaf emergence. In cherry (*Prunus cerasus* c.v. Montmorency) and *Malus hupehensis*, the rates of wax synthesis often did not compensate for the rapid increase in surface area and consequently wax deposits decreased during the course of development. In contrast, in dwarf bean and grape vine leaves, wax synthesis increased markedly during the exponential phase of growth, with the result that the amount of wax per unit area increased despite leaf expansion. Baker and Hunt[8] also found that the amount of wax on the adaxial and abaxial surfaces of the same leaf is age and species dependent. These results emphasize the danger inherent in generalized statements with regard to the effect of age on epicuticular wax development in dicotyledonous species.

Differences in the efficiency of cuticle penetration in various regions of the leaves of grasses or cereals may be explained in terms of differences in cuticule thickness or composition. Application of herbicides to the lamina base of grass leaves is known to improve herbicide performance and Coupland et al.[45] suggested that this results from reduced wax depositions and perhaps higher humidity in that region. Robertson and Kirkwood[46] found that increased absorption of [14C]asulam and [14C]isoproturon applied to the leaf base as opposed to the mid-blade region coincided with less well organized and less dense wax deposits in the former; differences in wax composition remain to be determined. This result is not unexpected in view of the manner of leaf development in *Gramineae* in which the oldest tissue of the blade occurs at the tip and the youngest tissue at the base. Absorption is greatly enhanced if the herbicide is applied to the inner surface of the leaf sheath in which epicuticular wax deposits are non-existent[47].

The difference in permeability of upper (adaxial) and lower (abaxial) leaf surfaces has been reported by several authors[30, 40, 48]. In dicotyledonous species the upper cuticle is generally thicker and often less permeable than the lower. However, the composition of epicuticular and possibly cuticular waxes differ for the two surfaces and this together with the differing distribution of stomata should be taken into account when comparing the rates of uptake.

Several environmental factors influence cuticle thickness and possibly composition, including light, temperature and humidity. Leaves developing under full sunlight form a thicker cuticle than those expanding in the shade, while high temperature and humidity conditions appear to be consistent with greater cuticle permeability[49]. Greenhouse-grown plants are generally believed to have thinner cuticles than those of field-grown plants. They are not necessarily more permeable, however, and it may be that the effects of the thinner cuticle are outweighed by differences imposed by the changed conditions[5].

It is clear that cuticle thickness cannot be considered in isolation from certain other factors that may influence cuticle permeability. For example, Reed and Tukey [50] studied the differences in cuticle permeability of isolated cuticles of leaves of Brussels sprouts and carnations grown in different temperatures and light intensities. Differences in permeability (P) were explained only partially by measured differences in partition coefficient (K), membrane thickness (Δx) and percentage of wax. The highest degree of correlation was found between flux (F) or P and the interaction $K/\Delta x$ (percentage of wax/100) derived from a modified Fick's law of diffusion, which explained around 79 per cent of the permeability differences between cuticles developed in the different environments. Reed and Tukey[50] concluded that no single measured cuticle parameter could be used to explain differences in cuticle permeability; the interaction of all of the examined parameters was important.

1.4.2 Temperature

The effect of temperature changes on diffusion rate across the cuticles in theory should be relatively small, at least within the physiological limits of plant growth. In practice, however, temperature changes may greatly increase cuticle penetration, especially near 30 °C, possibly reflecting phase changes in cuticular components at high temperatures resulting in decreased viscosity[5]. Such phase changes have been noted in the cuticles of Citrus auranticum[51].

The penetration of [14C]MCPA into cuticles stripped from the abaxial surface of leaves of Vicia faba was similarly increased by temperature (5–35 °C)[52]. These 'cuticles' incorporated the epidermal cells and it was suggested that absorption may be limited by a metabolically governed step which was increased by temperature ($Q_{10} > 2$). Treatment with metabolic inhibitors was largely inconclusive, however, and the possible role of the epidermal plasmalemma as a rate-limiting step remains to be confirmed. Phase changes may have been involved but were not determined.

1.4.3 Wax viscosity

The viscosity of cuticle waxes is reduced by organic solvents used to formulate certain pesticides, but it may not be possible to distinguish between effects on the solute and those on cuticle permeability[5]. In addition to effects on surface tension, surfactants may reduce wax viscosity, especially at higher concentrations, which may explain the enhanced rates of cuticle penetration of herbicides formulated with certain groups of surfactants. Surfactants are known to penetrate and possibly cross the cuticle. Price[5] has conjectured that these surfactants may change viscosity by mixing with amphipathic acids and alcohols of the wax. Amphipathic compounds, such as fatty acids and phospholipids, have both a hydrophobic and a polar portion.

1.4.4 Molecular radius

In view of the heterogeneous and molecular sieve characteristics of the cuticle, the rate of uptake tends to be more closely related to the minimum molecular radius than to the maximum molecular radius or molecular weight[5]. The situation is complicated, however, by the fact that the radius will be different in water than cuticle waxes since the molecular configuration of a solute depends on the nature of the solvent; a further complication concerns the binding of water to strongly polar compounds. This last characteristic may be involved in the relatively slow cuticle penetration of, for example, bromoxynil salt as opposed to the corresponding octanoate ester formulation[53]. The latter, being relatively non-polar, is believed to partition rapidly into the cuticle waxes in which the ester may be hydrolysed to the acid, thereafter crossing the cuticle as reduced molecular weight fragments of a relatively polar nature. An enzyme has been located in the cuticle that is capable of hydrolysing esters[5], and esterase activity on the surface of apple and marrow leaves has been reported[54].

1.4.5 Partition coefficients

The transmission of a solute across the cuticle is dependent on the efficiency with which it partitions into the cuticle from the external aqueous phase and again out of the cuticle to the inner aqueous phase[5]. An indication of the ability of the compound to partition is reflected in the partition coefficient, which normally is expressed as the logarithm of the ratio of the solubility in water and octanol. Compounds with a low partition coefficient are normally rapidly absorbed under high humidity conditions, supporting the view that there is an aqueous route traversing the cuticle. Compounds with a $\log P > 3$ generally have a high solubility in the waxes; they may be retained in the cuticle waxes and not partition into the aqueous phase on the inner regions of the cuticle. For example, the uptake of [^{14}C]MCPB through the cuticles of *Vicia faba* (*in vitro* or *in vivo*) is relatively slow compared with that of [^{14}C]MCPA due to selective retention of the former in the 'cuticle', probably reflecting the non-partitioning of the relatively lipophilic MCPB from the cuticular waxes[52]; this has also been suggested for 2,4-DB[55].

1.4.6 Effective area

Uptake into the tissue is proportional to the area available to the diffusing solute; in the case of a solute traversing an aqueous route this effective area may be very small. For example, isolated dewaxed cuticles of *Citrus auranticum* have an effective area for water diffusion of 0.08–0.24 per cent[56]. An increase in the effective area should increase uptake, but where surfactants are compared this may not occur. Other effects of the surfactants appear to obscure the area effect[57] and an increase in the latter may be less important than maximizing the solute activity in the surface deposit[5].

1.4.7 Formulation

Uptake generally proceeds as long as the spray deposit remains in the aqueous phase, loss of solvent inhibiting further uptake unless the deposit is rewetted by rain or dew. However, compounds with a high crystal lattice energy redissolve only with difficulty and careful formulation is required[5]. Solutes formulated in oil or other slow-evaporating solvents will tend to remain in solution longer and be less influenced by high humidity, dew or rain. Inclusion of a humectant that prevents the complete drying of the spray droplets may enhance the total uptake of a soluble compound while reducing the rate of uptake, even under low humidity conditions.

Surfactants or surface active agents are compounds that at low concentrations (0.01–0.1 per cent) can increase the activity of herbicidal sprays[58], largely reflecting reductions in surface or interfacial tension of the spray solution. Improvement in the wetting and spreading properties leads to greater retention and target coverage.

Three classes of surfactants have been identified:
1. Anionic surfactants that possess a long non-polar portion of the molecule with a hydrophilic ionized end portion.
2. Cationic compounds that have a long carbon chain confering hydrophobic qualities that counterbalance the hydrophilic charged nitrogen atom leading to separation of the ion at water–non-water phase boundaries. Since the amphipathic ion is positively charged, it cannot interact with cations in water or with cationic active ingredients.

3. Non-ionic surfactants are soluble in hydrocarbon and oily liquids, which facilitates the production of single-phase rather than dual-phase emulsifiable concentrates. Most non-ionic compounds are polyethylene oxide derivatives[59].

All surfactants contain hydrophilic and lipophilic groups and the balance between the opposing effects can be determined (HLB). Surfactants of low HLB value (e.g. Span 60, sorbitan monostearate) are relatively lipophilic while those of high HLB (e.g. Tween 20, polyoxyethylene sorbitan monolaurate) are relatively water soluble. This HLB system allows a degree of selection of surfactants for specific purposes.

The mechanism by which surfactants facilitate foliar entry and movement of herbicides has received little attention, perhaps reflecting the lack of suitable radiolabelled compounds. Studies with a number of radioactive surfactants (e.g. ^{14}C-Tween 20 [60], and ^{14}C-Tween 80[61, 62] have shown very limited foliar penetration, the acid and polyoxyethylene parts of the molecule being cleaved, most probably on the leaf surface. On the other hand, substantial amounts of surfactant uptake have been reported by other workers. Stolzenberg et al.[63] found a substantial amount of uptake from dilute solutions of Triton X-100 (OP-polyethoxylate, average oxyethylene (EO) content 9.5 per cent) containing homogeneous OPE6 or OPE9. Little translocation occurred but metabolism, mainly to polar products, occurred within the treated leaf after one day. Similarly, Anderson and Girling[64], using a colorimetric method of analysis, reported up to 80 per cent uptake within 48 h of four linear alcohol (C_{13}–C_{18}) polyethoxylates (E_{12}–E_{17}) into wheat leaves.

The properties of octylphenoxy surfactants (Triton X series) and their effects on foliar uptake of 2-deoxy-D-glucose (2D-glucose), atrazine and o,p^1-DDT by maize leaves has been investigated by Stevens and Bukovac[65]. This series of surfactants increased the uptake of these compounds, which had water solubilities ranging over six orders of magnitude. This resulted from two apparently distinct modes of action. The uptake of 2D-glucose (H_2O solubility 50 g/l) was enhanced by the surfactants maintaining the chemical in solution on the leaf surface. Stevens and Bukovac suggest that to maximize this effect, surfactants with long oxyethylene (EO) chains would probably be suitable for similar water-soluble active ingredients (a.i.). The uptake of atrazine (40 mg/l) and DDT (17 g/l) was related to penetration of the surfactants. The mode of action remains unknown but it appeared not to be attributable to solubility of the leaf surface waxes by the surfactants. They concluded that surfactants with short EO chains would be adjuvants of choice to maximize uptake of compounds of non-polar active ingredients. In the case of phloem-translocated herbicides, however, consideration of phytotoxicity suggests that surfactants with intermediate EO chain lengths may provide a preferable compromise between minimal adjuvant phytotoxicity and maximal uptake.

The foliar uptake, movement and metabolism of three non-ionic surfactants applied to a range of species has been studied[66] [$C_{12}E_8$] (a homogeneous [l-^{14}C]-labelled l-dodecanol octaethoxylate), $C_{18}E_{8.5}$ (an oligomeric mixture of [U-^{14}C] ethoxylate chain labelled l-octadecanol ethoxylates, with average EO content 8.5) and NPE 5.5 (an oligomeric mixture of [U-^{14}C]phenyl ring labelled nonyl phenol exthoxylates, with average EO content 5.5)]. Considerable quantities of these surfactants were absorbed, the amount and rate being dependent on both the plant and surfactant. In some plants penetration was very rapid, with as much as 90 per cent uptake occurring within 2 h of foliar application. The presence of microcrystalline deposits of epicuticular wax in certain species did not present a significant barrier to the penetration of the surfactants. Little translocation was evident, but metabolism occurred within the treatment region, the amount and rate varying with plant species.

Only a small proportion of the ^{14}C-labelled NPE 5.5 which penetrated into the leaves of barley or pea was associated with the epidermal layer (cuticle + epidermis)[67], suggesting that neither the epicuticular wax, cuticle or epidermis presented much of a barrier to penetration of NPE 5.5. The small amount recovered in the stripped epidermal layer could represent 'adsorption' or 'saturation' values. Conversely, in the case of ^{14}C-labelled $C_{18}E8.5$, over half the amount of radioactivity absorbed into the leaf consistently was retained in the epidermal layer, necessitating evaluation of the relative importance of the transport stages from surface to epidermis and epidermis to palisade/mesophyll.

Surfactants may also act as humectants, especially those of moderate to high hydrophilic/lipophilic balance (HLB), maintaining solute activity and facilitating the partitioning of solute into cuticle wax. In fact, activity of the solute in water is modified not by surfactants but by surfactant micelles, thus maintaining a high concentration of solute[57] which replenishes solute 'lost' by partition into the cuticle. The concentration of solute at the cuticle surface may be enhanced by accumulation of surfactant at this interface and by direct transfer from the micelles to the surfactant–cuticle interface[5].

To sum up, many of the factors influencing transmission of solute across the cuticle act on several parameters and no single parameter in the equation describing solute movement across the cuticle can be considered in isolation[5]. For example, organic solvents not only maintain a high external concentration, but may also influence the viscosity of the cutin waxes. In the case of the polar solutes that may accumulate in the leaf tissues or be translocated in the symplast, high concentrations may occur in the external aqueous phase, with relatively low concentrations in the cutin wax and inner surface of the cuticle. The concentration gradient across the cuticle may be enhanced by rapid translocation away from the inner regions of the cuticle, possibly as a result of 'sink' stimulation[68]. The effect of surfactants may be complex since they may modify the external concentration, the effective area, viscosity and tortuosity of the aqueous pores; additionally they may affect membrane permeability.

1.5 Environmental factors and cuticle permeability

The environmental conditions prevailing during plant growth may affect cuticular permeability since wax production varies according to ambient conditions. Changes in response to fluctuations in environmental conditions may be rapid since changes in surface fine structure are evident within 48 h[69].

The influence of environmental conditions on cuticle wax development and penetration of naphthylacetic acid (NAA) have been studied by Hunt and Baker[70]. Increased wax yields from the leaves of *Brassica oleracea* occurred with increase in radiant energy rate and reduced temperature and relative humidity; increased yields were also found from leaves of several species grown in 'dry' habitats. In particular, the effect of ambient conditions and soil moisture content (s.m.c.) on crystalline wax production on the leaves of *Pisum sativum* have been examined. The thickness of the wax deposit increased with a decrease in s.m.c. (40 per cent v. 100 per cent), irrespective of the prevailing ambient conditions; when the s.m.c. was held constant, however, wax production increased with a rise in vapour pressure deficit (v.p.d.), particularly on the adaxial surface. Scanning electron microscopy revealed that the size of the crystalline wax structure increased with a rise in v.p.d. while their density declined with an increase in s.m.c., the size and density of wax platelets increased on the adaxial surface of high-stress plants.

While wax composition varied only slightly with changes in these growth conditions, major differences were found between adaxial and abaxial surfaces; the former was composed mainly of primary alcohols (75–85 per cent) and the latter of hydrocarbons (70–80 per cent). Waxes from the abaxial surface of plants grown under high v.p.d. contained smaller amounts of hydrocarbons (45–50 per cent) and a higher proportion of primary alcohols (35 per cent).

Hunt & Baker[70] found that the rates of penetration of NAA into leaves of *P. sativum* correlated inversely with changes in the wax deposit. Thus penetration occurred most readily through the sparse wax deposits formed under a combination of low radiant energy flux and high soil moisture and humidity conditions.

In these studies the natural variations in epicuticular wax deposits which occurred during leaf development or which resulted from changes in growth conditions have been exploited. It is perhaps unavoidable that in addition to these desired changes variations in growth conditions may have affected other metabolic processes which in turn influence the processes of uptake and translocation. The findings do emphasize, however, the importance of pretreatment growth conditions in studies of uptake and movement, and underline the desirability of maintaining standard watering regions as well as controlled temperature, light and humidity conditions.

It is well known that pretreatment of plants with compounds which affect lipid biosynthesis may enhance the uptake of foliage-applied herbicides due to reduced wax deposition. This is exemplified by the effect of pretreatment of plants with sodium trichloroacetate (TCA), dalapon and carbamate herbicides; there is evidence that these compounds interfere with lipid biosynthesis[71].

More recently Hunt and Baker[70] found that EPTC treatment reduced the wax deposit on the adaxial surface of *Pisum sativum* leaves by 31 per cent and on the abaxial surface by 61 per cent compared to the controls. The wax from the abaxial surface of treated plants contained lower proportions of hydrocarbons and higher proportions of alkyl esters and alcohols. Scanning electron microscopy revealed marked changes in the distribution of the crystalline waxes. Uptake and translocation of NAA was greatly increased through both surfaces, the proportional increase of uptake by adaxial and abaxial surfaces being eightfold and threefold respectively. These results emphasize the importance that residues of the above herbicides may have on the uptake, translocation and activity of compounds applied subsequently, with possible implications for crop tolerance.

1.6 Sites of preferential entry

The existence and significance of sites of preferential entry is a somewhat controversial issue. These sites include stomata, trichomes, the cuticle associated with veins, anticlinical walls of epidermal cells and leaf bases. The role of stomata is still uncertain, though it is well known that absorption of foliage-applied chemicals is generally greater through the abaxial (lower) leaf surface of dicotyledons. This has been partly attributed to the greater number of stomata on the lower surface, though the confounding differences in the cuticle wax composition and structure make it difficult to identify the role of the stomata. The relative importance of the pore as opposed to guard cells as portals of entry has also to be resolved.

The mechanism of pore penetration should take into account the surface tension and contact angle of the droplet and the physicochemical characteristics of the pore wall. Penetration of the pore may be possible if the surface tension of the spray droplets is equal to or less than the surface tension of the plant surface or the minimum wall angle of the stomatal pore[72]. Assuming the critical surface tension of many cuticular surfaces to be approximately 30 mNm, this can be compared with surface tensions of 29.4 and 25.3 mNm of ethirimol formulated with 0.2 per cent Lissapol NX or Tergitol 7 respectively[5]. The mass movement of solute molecules from spray droplets into the leaves through the stomatal pores has been reported by Dybing and Currier[73], Schönherr and Bukovac[72] and Greene and Bukovac[74], but others are doubtful that sufficiently low surface tensions can be achieved using current surfactants. Even if entry via the pore does occur it would still be necessary for the solute molecule to penetrate the relatively thin cuticle which encloses the substomatal cavity[29], though the relatively high humidity of the cavity must facilitate penetration of that cuticle.

Other workers believe that preferential penetration may occur via the guard cells and accessory cells (e.g. Sargent and Blackman[75]), and in such cases stomatal uptake would be determined by the size and distribution of the stomata rather than the dimension of the stomatal aperture[70]. The thickness of the wax layer in the region of the stomata may differ from that of the surrounding tissues[8], and differences in the form and distribution of crystalline deposits around the stomatal complex are often found on plants that have glaucous surfaces[76–78].

It has been suggested that preferential penetration of the cuticle over the guard cell, basal cells of trichomes, anticlinal walls of epidermal cells, etc., may reflect the relatively high numbers of ectodesmata (or 'ectocythodes') reputed to occur at those sites[79]. While doubt has been cast on the existence of these controversial 'structures' by Schönherr and Bukovac[80], these authors believe that areas do exist in the cuticle that are preferentially permeable to polar compounds. There is some evidence to support the view that hydrophilic compounds preferentially occupy those areas of the cuticle that are characterized by polysaccharide microfibrils. Hoch[81] observed localized precipitation of silver thiocarbohydrazide following application of silver nitrate to the surface of cuticles of *Malus pumila* floating on a solution of thiocarbohydrazide. It must be noted, however, that 'ectodesmata' do not pass through the cuticle, but only through the epidermal cell wall into the cuticle. Thus they cannot provide a route for the initial stages of entry of hydrophilic compounds[35].

Localized areas of retention and/or penetration appear to exist in 'cuticles' (including the epidermal cells) isolated from the abaxial surface of *Vicia faba* treated with Rhodamine B dye[52]. The dye was found to be preferentially associated with the stomata, trichomes and cuticle above the anticlinical walls, and a similar pattern was obtained by microautoradiography of isolated 'cuticles' treated with [14C]MCPA. Incorporation of Span and Tween surfactants to produce a lipophilic HLB (68.6) resulted in a more general distribution of [14C]-label over the cuticle surface.

1.7 Tissue absorption

Cuticle penetration is the first step in the overall process of leaf absorption, part of which may involve an energy-requiring step. The conditions that should be satisfied in determining whether an active mechanism is involved in absorption have been listed by Donaldson *et al.*[82]:

1. The temperature coefficient (Q_{10}) is > 2.
2. The process is oxygen requiring.
3. Rate of absorption is hyperbolic rather than linear.
4. Compounds of similar structure exhibit competitive inhibition.
5. Absorption is reduced or blocked by metabolic inhibitors such as 2,4-dinitrophenol.
6. Accumulation takes place against a concentration gradient, i.e. the concentration is higher within the tissue than in the solution.

Several authors have demonstrated that the initial stages of absorption are rapid, possibly being controlled by a passive mechanism, while the later phase appears to involve a metabolic component. For example, the foliar uptake of asulam by bracken (*Pteridium aquilinum*)[83] and dalapon by *Lemna minor*[84] exemplify this phenomenon. Studies of the kinetics of uptake of the phenoxyacetic acids have shown the presence of two distinct phase of uptake[35]. The first phase is unaffected by metabolic inhibitors and presumably reflects simple diffusion through the cuticle and into the free space of the leaf tissue; the second phase is sensitive to metabolic inhibitors and probably represents an energy-dependent movement across the plasmalemma (cell outer membrane) into the cell[85].

While the cell wall is generally regarded as being freely permeable, there is a possibility that some of the herbicide that penetrates the cuticle may be immobilized in the apoplast (e.g. Donaldson *et al*.82) or subject to electrostatic repulsion. Both phenomena require to be substantiated and their importance is difficult to assess.

The uptake of herbicides into the symplast has been studied using leaf slices[86] or cell suspension cultures (e.g. Rubery and Sheldrake[87], Kurkjian *et al*.[88]). Examination of the uptake of phenoxyacetic acids into the symplast revealed that as the herbicide concentration in the medium increased so the rate of uptake increased to a maximum. After this point further increases in concentration were ineffective[89, 90], suggesting the operation of a saturable membrane carrier or an energy requirement, or both.

The involvement of a metabolically governed step in the uptake process is well documented[35] but the nature of its involvement is a matter of controversy. It is becoming more widely accepted that energy is required to maintain a pH gradient between the cell wall and cytoplasm and that herbicide entry requires the existence of such a gradient. There is evidence of an energy-requiring proton pump in the plasmalemma of plant cells that is responsible for the proton gradient across the membrane[91, 92].

The possible mechanisms by which the uptake of the phenoxyacetic acids can be linked to this proton gradient have been reviewed by Pillmoor and Gaunt[35]. The diffusive uptake model proposed by Rubery and Sheldrake[93] is based upon the pH of the medium. At low pH, uptake is much greater than at pH values approaching neutrality, and it is proposed that a phenoxyacetic acid can only diffuse across the plasma membrane in the un-ionized, lipophilic form; no passive diffusion of the dissociated anion can occur. Being weak acids, the degree of dissociation of these compounds depends on the pH of the medium. If the pH of the cell wall is 5–6 and that of the cytoplasm is 7, then the degree of diffusion in the former would be relatively less than in the cytoplasm. It is envisaged that, under these circumstances, diffusion of undissociated acid will take place across the plasmalemma into the cytoplasm and down a concentration gradient, until the concentration of undissociated herbicide in the cell wall and cytoplasm is the same. The energy requirement of the uptake process is involved partly in maintaining the pH gradient across the plasmalemma and partly in keeping the cytoplasm pH constant[92].

Other work suggested that carrier proteins in the membrane may be involved in the movement of auxins across the cell membrane[87, 89, 90, 94, 95]. It is proposed that initially the anions of IAA and 2,4-D could be transported into cells by means of an electroneutral proton symporter, the process involving gradients of both anions and protons across the plasmalemma. The importance of this carrier system compared with diffusion of undissociated auxin is dependent upon the pH of the external medium. At a pH of 4.0, optimum for carrier-mediated uptake of 2,4-D, the ratio of carrier-mediated transport to entry by diffusion is 1:6.7; as the pH is reduced so this ratio increases markedly. At pH 5, however, the ratio decreased to 1:2.5, suggesting that in intact leaves both pathways may be important under normal conditions. The carrier may later be involved in the efflux of auxins and a dynamic equilibrium may be reached when the rate of entry by diffusion of the un-ionized auxin is equal to the rate of efflux by the anion carrier. The existence of a second auxin carrier has been proposed[94, 95] that mediates the electrogenic export of auxin anion and is sensitive to trichlorobenzoic acid.

Any process that affects the gradient of herbicide across the plasmalemma will affect the rate and extent of cuticle penetration[35]. These processes include movement of the herbicide to other cells within the leaf, export from the leaf or movement into the vacuole or cell organelles. Pillmoor and Gaunt[35] have applied the model of Rubery and Sheldrake[87] to the partition of phenoxyacetic acid between the cytoplasm, vacuole and organelles. They suggest that distribution will depend largly upon the pH in each compartment. The vacuole (pH of, say, 5) might be expected to accommodate less phenoxyacetic acid than the cytoplasm, but because of the relatively large volume of the vacuole the total amount of herbicide may be greater than the cytoplasm at equilibrium. The relatively high pH of active mitochondria and chloroplasts (e.g. stroma around pH 8) could result in relatively high concentrations of herbicide accumulating in these organelles. This underlines the importance of considering the levels of organelle accumulation in mode-of-action studies, since apparently low tissue levels may belie the situation within the organelles which may represent the ultimate sites of action.

Pillmoor and Gaunt[35] also point out that the uptake of phenoxyacetic acids from the cell wall (and thus the cuticle surface) will also be enhanced by any metabolism of the herbicide that reduces cytoplasmic concentration of by non-covalent binding of herbicide to protein or other macromolecules in the cytoplasm. It is also possible for auxins to increase their own uptake (1) by affecting the permeability of the plasmalemma[96, 97], which could potentially increase their own uptake[98], (2) by stimulating a plasmalemma-bound proton pump[35] and (3) by developing new 'sinks' (regions of high metabolic activity and thus sugar demand). This would increase the translocation of assimilates and thus herbicide to these areas, leading indirectly to enhanced uptake.

The increased activity of certain herbicides when formulated with certain ammonium compounds (e.g. ammonium thiocyanate) may be due to increased absorption (e.g. asulam[99]), perhaps reflecting increased membrane permeability[30]. The incorporation of 0.1–2.5 per cent surfactant further increased the effect of ammonium sulphate (0.1–2.5 per cent w/v) on glyphosate activity[100], presumably reflecting enhanced cuticle penetration of the adjuvant to its site of action.

1.8 Translocation from the region of absorption

Once within the leaf tissues, translocation of systemic compounds may take place by apoplastic or symplastic routes; at herbicidal doses contact herbicides are not normally translocated to any degree. Herbicides that are translocated primarily in the xylem vessels or tracheids are said to exhibit an apoplastic pattern of movement, while those that are translocated in the sieve tubes of the phloem demonstrate a symplastic pattern. Some compounds can move more freely in either the xylem or phloem and are termed ambimobile (e.g. dalapon). 'Psuedo-apoplastic' compounds (e.g. the nematicide oxamyl) are able to freely exchange between the apoplast and symplast, but they are not retained in the phloem to any extent, usually being carried away in the transpiration stream [101].

Substances that move apoplastically are transported upwards in the leaves, accumulating at the tips in monocotyledons and around the leaf margins in the case of dicotyledons. Herbicides moving in the symplast show a pattern of transloca-tion that is similar to that of assimilates; they can move upwards or downwards, generally accumulating in tissues of high assimilate demand ('sinks'). This movement of phloem-translocated herbicides follows a 'source' to 'sink' pattern.

Consideration of the detailed nature of translocation mechanisms is outside the scope of this review. It is perhaps sufficient to state that the whole question of the nature of phloem transport is controversial and that several concepts of movement have been proposed[102, 103]. However, the indirect effect of translocation on the kinetics of cuticle penetration and symplast absorption is relevant.

Any mechanism that removes herbicide molecules from the absorption region must enhance the concentration gradient across the cuticle. Indeed, positive correlations between absorption and translocation have been reported for several compounds including MCPA[58] and asulam[83]. Clearly absorption is a prerequisite for translocation, but the converse may also be true; enhanced translocation may indirectly result in an increased rate of cuticle penetration. Thus the influence of environmental factors such as light and temperature may enhance photosynthesis and the phloem transport of assimilates together with any associated herbicide molecules; uptake may be indirectly affected.

In addition, any factor that induces sink activity will enhance source–sink translocation, indirectly increasing absorption. For example, induction of sink activity by certain growth regulator treatments resulted in enhanced absorption and translocation of ^{14}C-labelled herbicides[53]. It would appear that the induction of assimilate and herbicide flow to the sinks indirectly influenced the rate of absorption and cuticle penetration.

The binding or conjugation of herbicide to macromolecules at non-metabolically active sites may enhance the diffusion gradient of penetrating molecules. Indeed, the possibility that plants can form 'emergency metabolites' by rapidly forming conjugates with herbicidal compounds such as the phenoxyalkanoic acid compounds has been discussed by Pillmoor and Gaunt[35]. Such conjugates may act as a temporary reservoir of potentially active herbicide which can be released later on breakdown of the conjugate.

Phytotoxicity and the effects on leaf cell membrane permeability may be important factors determining the efficiency of phloem translocation of assimilates and herbicide. Photosynthetic inhibitors such as linuron or prometryne rapidly increase membrane permeability, as does paraquat. Conversely, the slow-acting or hormone-type herbicides such as picloram or 2,4-D affect membrane permeability more slowly, if at all[104]. It must be presumed that phytotoxic effects resulting in increased membrane permeability may result, at least temporarily, in enhanced absorption, particularly if membrane transport was a rate-limiting factor. However, translocation of herbicides that rapidly affect membrane integrity does not normally occur, though the effects are concentration dependent.

The effect of herbicides on the energy metabolism of leaf cells may influence the efficiency of translocation, since at least the 'loading' of minor veins is believed to be ATP dependent. The action of uncouplers or inhibitors of oxidative phosphorylation[105] or photophosphorylation[106] may reduce the availability of ATP for energy-dependent processes. Many contact herbicides act as inhibitory uncouplers, and this, together with the effects on membrane permeability, may adversely affect their phloem translocation. The effect of such compounds on assimilate production and translocation is an additional and related factor. Inhibition of pigment synthesis by 'bleaching' herbicides such as norflurazon[107] or diflufenican[108] may indirectly influence the extent and rate of translocation and possibly absorption.

An indication of the loss of herbicide by transformation in sunlight or ultraviolet light can be obtained by measuring the disappearance of the starting material. For example, the loss of ^{14}C-labelled material applied to a series of glass coverslip fragments can be determined by radioassay of the glass fragments at intervals during the course of an uptake study. It is likely that this assay will result in an overassessment of the loss, since only a portion of the dose applied to leaves remains on the surface. This assessment is best regarded as a measure of the potential loss rather than actual loss of the ^{14}C-label under the particular experimental conditions.

1.9 Herbicide metabolism

While the topic of this review does not strictly encompass herbicide metabolism, it is relevant to mention its importance in the context of interpretation of data concerning uptake and movement. In considering this aspect of the herbicide mode of action, many workers have used a range of techniques that are now fairly commonplace in their application in plant physiology.

Autoradiography of the ^{14}C-treated plants enables a qualitative view of absorption and distribution of the label in the various regions. However, the distribution of the label may not necessarily be indicative of the herbicide *per se*. Quantitative radioassay of uptake and movement can be carried out by liquid scintillation counting. At harvest, the treated plants are divided into various regions such as treated leaves, roots and untreated shoots. The dry weight of the tissues is measured prior to oxygen combustion of each region and the [^{14}C] formed is trapped in a suitable absorbent/scintillant which is radioassayed. The herbicide residues on the surface of the treated leaves can be determined by radioassay of water and chloroform washes respectively or by 'stripping' the surface with cellulose acetate[5]. Again, this quantitative assay evaluates only the distribution of the ^{14}C-label and not necessarily the herbicide active ingredient.

Degradation of herbicide molecules may occur on the leaf surface or within the plant *en route* to the ultimate sites of action. A number of reviews have been concerned with herbicide metabolism[49, 109–111] and the relationship of metabolism studies to the herbicide mode of action[112]. It is sufficient here to emphasize that, at the very least, a check of the fate of the ^{14}C-labelled herbicide should be made under the same time scale and experimental conditions as the uptake/translocation studies. Extraction, cleanup and separation by chromatography is necessary to isolate the herbicide from its metabolites. Paper or thin-layer chromatograms may be subjected to radiochromatogram assay or to autoradiography to determine the location of the radio label. If the existence of radiolabelled metabolites is indicated, the quantity of these relative to the herbicide can be determined. The chromatograms can be cut into strips containing the herbicide or its labelled metabolites for quantitative assay by liquid scintillation counting. If separation was carried out by gas–liquid chromatography (g.l.c.) then the gas effluent can be monitored by fraction collection in a suitable absorbent scintillant.

Generally such metabolism checks are particularly necessary in the case of more long-term studies and often form an integral part of studies concerned with an understanding of the basis of herbicide selectivity. Some of the technical difficulties that may be encountered in examining herbicide metabolism have been considered by Pillmoor and Gaunt[35].

1.10 Techniques for studying the retention, penetration and translocation of foliage-applied herbicides

The experimental methods used in the investigation of the mode of action of foliage-applied herbicides have been reviewed recently[113, 114], with particular reference to the removal and assessment of foliar deposits using cellulose acetate film stripping[67], penetration of surfactants using epidermal techniques[115, 116] and sample preparation for liquid scintillation counting[117].

1.11 Conclusions

Much effort has been devoted to the study of the physicochemical properties of leaf surfaces and the processes involved in penetration of foliage-applied herbicides. The role of the epicuticular and cuticular waxes is relatively well understood, at least in the case of certain species. However, the pathways of transcuticle movement, sites of preferential penetration and the nature of the rate-limiting processes involved in absorption are still controversial.

The results of studies concerned with movement of foliage-applied herbicide tend to support the view that the pathways and mechanisms of distribution of phloem-mobile herbicides are similar to those of assimilates. However, little evidence is currently available on the mechanisms of 'loading' or short- and long-distance transport of herbicidal compounds. The efficiency of these processes may indirectly influence cuticle penetration and absorption of herbicides.

1.12 References

1. P.J. Holloway, 'Structure and histochemistry of plant cuticular membranes: An overview' in *The Plant Cuticle*, ed. by D.F. Cutler, K.L. Alvin and C.E. Price, Linnaeus Society Symposium Series No. 10, Academic Press, London (1982), pp. 1–32.
2. R.F. Norris and M.J. Bukovac, 'Structure of the pear leaf cuticle with special reference to cuticular penetration', *Am. J. Bot.*, **55**, 975–83 (1968).
3. N.D. Hallam, 'Fine structure of the leaf cuticle and the origin of leaf waxes', in *The Plant Cuticle*, ed. by D.F. Cutler, K.L. Alvin and C.E. Price, Linnaeus Society Symposium Series, No. 10, Academic Press, London (1982), pp. 197–214.
4. H.C. Hoch, 'Penetration of chemicals into the *Malus* leaf cuticle. An ultrastructural analysis', *Planta*, **147**, 186–95 (1979).
5. C.E. Price, 'A review of the factors influencing the penetration of pesticides through plant leaves', in *The Plant Cuticle*, ed. by D.F. Cutler, K.L. Alvin and C.E. Price, Linnaeus Society Symposium Series No. 10, Academic Press, London (1982), pp. 237–52.
6. P.J. Holloway, 'The chemical constitution of plant cutins', in *The Plant Cuticle* ed. by D.F. Cutler, K.L. Alvin and C.E. Price, Linnaeus Society Symposium Series No. 10, Academic Press, London (1982), pp. 45–85.
7. E.A. Baker, 'Chemistry and morphology of plant epicuticular waxes', in *The Plant Cuticle* ed. by D.F. Cutler, K.L. Alvin and C.E. Price, Linnaeus Society Symposium Series No. 10, Academic Press, London (1982), pp. 139–65.
8. E.A. Baker and Grace M. Hunt, 'Development changes in leaf epicuticular waxes in relation to foliar penetration', *New Phytol.*, **88**, 731–47 (1981).
9. C.E. Jeffree, E.A. Baker and P.J. Holloway, 'Ultrastructure and recrystallisation of plant epicuticular waxes', *New Phytol.*, **75**, 539–49 (1975).
10. C.E. Jeffree, E.A. Baker and P.J. Holloway, 'Origins of the fine structure of plant epicuticular waxes', in *Microbiology of Aerial Plant Surfaces*, ed. by C.H. Dickinson and T.F. Preece, Academic Press, London (1976), pp. 119–58.
11. T.C. Chambers, I.M. Ritchie and M.A. Booth, 'Chemical models for plant wax morphogenesis', *New Phytol.*, **77**, 43–9 (1976).
12. A.M.S. Silva Fernandes, 'Studies on plant cuticle. VIII. Surface waxes in relation to water-repellancy', *Ann. Appl. Biol.*, **56**, 297–304 (1965).

13. R.F. Norris and M.J. Bukovac, 'Influence of cuticular waxes on penetration of pear leaf cuticle by 1-napthhaleneacetic acid', *Pestic. Sci.*, **3**, 705–8 (1972).
14. J. Schönherr, 'Naphthalene acetic acid permeability of *Citrus* leaf cuticle', *Biochemie und Physiologie der Pflanzen*, **171**, 25–31 (1976).
15. G.A. Matthews and R.P. Garnett, 'Herbicide application', in *Recent Advances in Weed Research*, ed. by W.W. Fletcher, CAB (1983), pp. 121–40.
16. W.B. Ennis and R.E. Williamson, 'Influence of drop size on effectiveness of low-volume herbicidal spray', *Weeds*, **11**, 67–72 (1983).
17. K.S. McKinley, S.A. Brandt, P. Mosse and R. Ashford, 'Droplet size and phytotoxicity of herbicides', *Weed Sci.*, **20**, 450–2 (1972).
18. G.D. Crabtree and M.J. Bukovac, 'Studies on low volume application of plant growth substances. Part I. Ethylene production, induced by 1-naphthylacetic acid, as a means of evaluating spray parameters', *Pestic. Sci.*, **11**, 43–52 (1980).
19. F.B. Hess, D.E. Bayer and R.H. Falk, 'Herbicide dispersal patterns. 1. As a function of leaf surface', *Weed Sci.*, **22**, 394–491 (1974).
20. C.R. Merritt, 'The influence of application variables on the biological performance of foliage-applied herbicides', *BCPC Monogr.*, **24**, 35–43 (1980).
21. G.W. Cussans and W.A. Taylor, 'A review of research on controlled drop application at the ARC Weed Research Organisation', *Proceedings of 1976 British Crop Protection Conference on Weeds* (1976), pp. 885–94.
22. W.A. Taylor and C.R. Merritt, 'Some biological aspects of very low and ultra-low volume application of MCPA', *BCPC Monogr.* **11**, 53–8 (1974).
23. C.R. Merritt and W.A. Taylor, 'Glasshouse trials with controlled droplet application of some foliage-applied herbicides', *Weed Res.*, **17**, 241–5 (1977).
24. M.S. Sokolov, V.S. Roskin, B.A. Kryzhro, B.P. Stickozov and V.V. Izubenko, 'The application of herbicides by low and ultra-low volume methods', *Khim. Sel'skom Khozyaistre*, **12**, 300–3 (1974).
25. P. Jegatheeswaran, 'Factors concerning the penetration and distribution of drops in low growing crops', *BCPC Monogr.*, **24**, 91–9 (1978).
26. B.D. Ripley and L.V. Edgington, 'Internal and external plant residues and relationships to activity of pesticides', *Proceedings of 10th International Congress on Plant Protection*, Vol. 2 (Brighton, 1983), pp. 545–53.
27. C.A. Hart, 'Use of the scanning electron microscope and cathodoluminescence in studying the application of pesticides to plants', *Pestic. Sci.*, **10**, 341–57 (1979).
28. H.M. Hull, 'Leaf structure as related to absorption of pesticides and other compounds', *Residue Rev.*, **31**, 1–155 (1970).
29. J.T. Martin and B.E. Juniper, *The Cuticles of Plants*, Arnold, London (1970).
30. R.C. Kirkwood, 'Some criteria determining penetration and translocation of foliage-applied herbicide', in *Herbicides and Fungicides, Factors Affecting Their Activity*, ed. by N.R. McFarlane, Special publication No. 29, The Chemical Society, London (1977), pp. 67–80.
31. L.A. Norris and V.H. Freed, 'The absorption, translocation and metabolism characteristics of 4-(2,4-dichlorophenoxy)butyric acid in big leaf maple', *Weed Res.*, **6**, 283–91 (1966).
32. R.G. Richardson 'A review of foliar absorption and translocation of 2,4–D and 2,4,5–T', *Weed Sci.* **25**, 378–90 (1977).
33. S.S. Szabo and K.P. Buckholtz, 'Penetration of living and non-living surfaces by 2,4–D as influenced by ionic additives', *Weeds*, **9**, 177–84 (1961).
34. J.R. Baur, R.W. Bovey and I. Riley, 'Effect of pH on foliar uptake of $2,4,5-T-1^{14}C$', *Weed Sci.*, **22**, 481–6 (1974).
35. J.B. Pillmoor and J.K. Gaunt, 'The behaviour and mode of action of the phenoxyacetic acids in plants', in *Progress in Pesticides Biochemistry*, ed. by D.H. Hutson and T.R. Roberts, Vol. 1, Wiley (1981), pp. 147–217.
36. P.S. Nobel, *Biophysical Plant Physiology*, W.H. Freeman, San Francisco (1970).
37. R.H. Scheiferstein and W.E. Loomis, 'Wax deposits on leaf surfaces. *Plant Physiol. Lancaster*, **31**, 240–7 (1956).
38. H.M. Hull, 'Leaf structure as related to penetration of organic substance', in *Absorption and Translocation of Organic Substances in Plants*, (ed. by J. Hacskaylo, 7th Annual Symposium of the American Society of Plant Physiology, S. Sect. (1964), pp. 45–93.
39. R.F. Norris, 'Penetration of 2,4–D in relation to cuticle thickness', *Am. J. Bot.*, **61**, 74–9 (1974).

40. M.J. Bukovac, J.A. Flore and E.A. Baker, 'Peach leaf surfaces: Changes in wettability, retention, cuticular permeability and epicuticular wax chemistry during expansion with special reference to spray application', *J. Am. Soc. Hort. Sci.*, **104**, 611–17 (1979).

41. A.P. Tulloch 'Composition of leaf surface waxes of *Triticum* species: Variation with age and tissue', *Phytochem.*, **12**, 2225–32 (1973).

42. E.A. Baker, J. Procopiou and G.M. Hunt, 'The cuticle of *Citrus* species. Composition of leaf and fruit waxes', *J. Sci. Food Agric.*, **26**, 1093–101 (1975).

43. B. Freeman, L.G. Albrigo and R.H. Biggs, 'Ultrastructure and chemistry of cuticular waxes of developing *Citrus* leaves and fruits', *J. Am. Soc. Hort. Sci.*, **104**, 801–8 (1979).

44. B. Freeman, L.G. Albrigo and R.H. Biggs, 'Cuticular waxes of developing leaves and fruits of blueberry *Vaccinium ashei* Reade cv. bluegem', *J. Am. Soc. Hort. Sci.*, **104**, 398–403 (1979).

45. D. Coupland, W.A. Taylor and J.C. Caseley, 'The effect of site of application on the performance of glyphosate on *Agropyron repens* and barban, benzoylprop-ethyl and difenzoquat on *Avena fatua*', *Weed Res.*, **18**, 123–8 (1978).

46. J. Robertson and R.C. Kirkwood, 'Herbicide uptake and translocation in grasses: Effect of site of application', in *Proceedings of 10th International Congress on Plant Protection* (Brighton, 1983), p. 575.

47. J. Robertson, N.L. Speight and R.C. Kirkwood, 'Uptake and translocation of [^{14}C] asulam and [^{14}C] isoproturon in grasses: Effect of site of application', submitted for publication in *Weed. Res.*

48. M.G. King and S.R. Radosevich, 'Tanoak (*Lithocarpus densiflorus*) leaf surface characteristics and absorption of triclopyr', *Weed Sci.*, **27**, 599–604 (1979).

49. W.W. Fletcher and R.C. Kirkwood, *Herbicides and Plant Growth Regulators*, Granada Publishing Company, London (1982), 408pp.

50. D.W. Reed and H.B. Tukey Jr, 'Permeability of Brussels sprouts and carnation cuticles from leaves developed in different temperature and light intensities', in *The Plant Cuticle*, ed. by D.F. Cutler, K.L. Alvin and C.F. Price, Linnaeus Society Symposium Series No. 10, Academic Press, London (1982), pp. 1–32.

51. K. Eckl and H. Gruler, 'Phase transitions in plant cuticles', *Planta*, **150**, 102–13 (1980).

52. R.C. Kirkwood, Irene McKay and R. Livingstone 'The use of model systems to study the cuticular penetration of ^{14}C-MCPA and ^{14}C-MCPB, in *The Plant Cuticle*, ed. by D.F. Cutler, K.L. Alvin and C.E. Price, Linnaeus Society Symposium Series No. 10, Academic Press, London (1982), pp. 253–266.

53. R.C. Kirkwood, 'The uptake and translocation of foliar-applied herbicides using an explant system', in *Advances in Pesticide Science*, ed. by H. Geissbuhler, Part 3, Pergamon, Oxford (1979), pp. 410–15.

54. D.R. Clifford, E.C. Hislop and C. Shellis, 'Effects of fungal and plant esterases on the fungicidal activity of nitrophenyl esters', *Pestic. Sci.*, **8**, 13–22 (1977).

55. M.A. Loos, In *Herbicides: Chemistry, Degradation and Mode of Action*, ed. by P.C. Kearney and D.D. Kaufman, Vol. 1, Marcel Dekker, Inc., New York (1975), pp. 1–128.

56. J. Schönherr, 'Transcuticular movement of xenobiotics', in *Advances in Pesticide Science*, ed. by H. Geissbuhler, Pergamon, Oxford (1978), pp. 392–400.

57. C.E. Price, 'Penetration and translocation of herbicides and fungicides in plants', in *Herbicides and Fungicides — Factors Affecting Their Activity*, ed. by N.R. McFarlane, Chemical Society Special Publication No. 29, The Chemical Society, London (1976), pp. 42–66.

58. G.S. Hartley and I.J. Graham-Bryce, *Physical Principles of Pesticide Behaviour*, Vol. 1, Academic Press, London (1980).

59. K.A. Hassall, *The Chemistry of Pesticides*, The MacMillan Press Limited, London (1982).

60. L.W. Smith and C.L. Foy, 'Penetration and distribution studies in bean, cotton and barley from foliar and root applications of Tween–20–^{14}C, fatty acid and xyethylene labelled', *J. Agric. Food Chem.*, **14**, 117–22 (1966).

61. C.G. McWhorter, 'The physiological effects of adjuvants on plants', in *Weed Physiology*, Vol. II: *Herbicide Physiology*, ed. by S.O. Duke, CRC Press Inc., Boca Raton (1985), pp. 141–58.

62. Y. Sugimura and T. Takeno, 'Behaviour of polyoxyethylene sorbitan ^{14}C-monooleate in tobacco and kidney bean leaves', *Pestic. Sci.*, **10**, 233–9 (1985).

63. G.E. Stolzenberg, P.A. Olson, R.G. Zaylskie and E.R. Mansager, 'Behaviour and fate of ethoxylated alkylphenol nonionic surfactant in barley plants', *J. Agric. Food Chem.*, **30**, 637–44 (1982).

64. N.H. Anderson and J. Girling, 'The uptake of surfactants into wheat', *Pestic. Sci.*, **14**, 399–404 (1983).
65. P.J.G. Stevens and M.J. Bukovac, 'Properties of octylphenoxy surfactants and their effects on foliar uptake', *Proceedings of British Crop Protection Conference on Weeds* (1985), pp. 305–16.
66. P.J. Holloway and Dawn Silcox, 'Behaviour of three nonionic surfactants following foliar application', *Proceedings of British Crop Protection Conference on Weeds* (1985), pp. 297–302.
67. D. Silcox and P.J. Holloway, 'A simple method for the removal and assessment of foliar deposits of agrochemicals using acetate film stripping', *Aspects Appl. Biol.*, **11**, 13–17 (1986).
68. R.C. Kirkwood, J. Dalziel, A. Matlib and L. Somerville, 'The role of translocation in selectivity of herbicides with reference to MCPA and MCPB', *Pestic. Sci.*, **3**, 307–21 (1972).
69. E.A. Baker, 'The influence of environment on leaf wax development in *Brassica oleracea* var. *gemmifera*', *New Phytol.*, **73**, 953–66 (1974).
70. G.M. Hunt and E.A. Baker, 'Developmental and environmental variations in plant epicuticular waxes: Some effects on the penetration of naphthylacetic acid', in *The Plant Cuticle*, ed. by D.F. Cutler, K.L. Alvin and C.E. Price, Linnaeus Society Symposium Series no. 10, Academic Press, London (1982), pp. 279–92.
71. G.G. Still, D.G. Davis and G.L. Zander, 'Plant epicuticular lipids: Alteration by herbicidal carbamates', *Plant Physiol. Baltimore*, **46**, 307–14 (1970).
72. J. Schönherr and M.J. Bukovac, 'Penetration of stomata by liquids. Dependence on surface tension, wettability and stomatal morphology', *Plant Physiol. Lancaster*, **49**, 813–19 (1972).
73. C.D. Dybing and H.B. Currier, 'Foliar penetration by chemicals', *Plant Physiol. Lancaster*, **36**, 169–74 (1961).
74. D.W. Greene and M.J. Bukovac, 'Penetration of naphthaleneacetic acid into pear (*Pyrus communis* L.) leaves. *Plant and Cell Physiol.*, **13**, 321–30 (1974).
75. J.A. Sargent and G.E. Blackman, 'Studies on foliar penetration. 1. Factors controlling the entry of 2,4–dichlorophenoxyacetic acid', *J. Exp. Bot.*, **13**, 358–68 (1962).
76. C.E. Jeffree, R.P.C. Johnson and P.G. Jarvis, 'Epicuticular wax in the stomatal antichamber of Stitka spruce and its effects on the diffusion of water vapour and carbon dioxide', *Planta*, **98**, 1–10 (1971).
77. J.A. Hanover and D.A Reicosky, 'Surface wax deposits on foliage of *Picea pungens* and other conifers', *Am. J. Bot.*, **58**, 681–7 (1971).
78. P.J. Holloway and E.A. Baker, 'The aerial surface of high plants, in *Principles and Techniques of Scanning Electron Microscopy*, ed. by M.A. Hayat, Vol. 1, Van Nostrand Reinhold Company, New York (1974), pp. 181–205.
79. W. Franke, 'The entry of residues into plants via ectodesmata (ectocythodes)', *Residue Rev.*, **38**, 81–115 (1971).
80. J. Schönherr and M.J. Bukovac, 'Preferential polar pathways in the cuticle and their relationship to ectodesmata', *Planta*, **92**, 189–207 (1970).
81. H.C. Hoch, 'Penetration of chemicals into the *Malus* leaf cuticle. An ultrastructural analysis', *Planta*, **147**, 186–95 (1979).
82. T.W. Donaldson, D.E. Bayer and O.A. Leonard, 'Absorption of 2,4-dichlorophenoxyacetic acid and 3-(p-chlorphenyl)-1,1-dimethylurea (monuron) by barley roots', *Plant Physiol. Lancaster*, **52**, 638–45 (1973).
83. P. Veerasekaran, R.C. Kirkwood and W.W. Fletcher, 'Studies on the mode of action of asulam in bracken (*Pteridium aquilinum*(L.)Kuhn). 1. Absorption and translocation of [^{14}C] asulam', *Weed Res.*, **17**, 33–9 (1977).
84. R. Prasad and G.E. Blackman, 'Studies on the physiological action of 2,2-dichloropropionic acid. III. Factors affecting the level of accumulation and mode of action', *J. Exp. Bot.*, **16**, 545–68 (1965).
85. J. Van Overbeek, 'Absorption and translocation of plant regulators', *Ann. Rev. Plant Physiol.*, **7**, 355–72 (1956).
86. R.C. Smith and E. Epstein, 'Ion absorption by shoot tissue: Technique and first findings with excised leaf tissue of corn', *Plant Physiol. Lancaster*, **39**, 338–41 (1964).
87. P.H. Rubery and A.R. Sheldrake, 'Effect of pH and surface charge on cell uptake of auxin', *Nature, New Biol.*, **244**, 285–8 (1973).
88. A. Kurkdjian, J.J. Leguay and J. Guern, 'The influence of fusicoccin on the control of cell division by auxins', *Plant Physiol. Lancaster*, **64**, 1053–7 (1979).

89. P.H. Rubery, 'The specificity of carrier-mediated auxin transport by suspension-cultured crown gall cells', *Planta*, **135**, 275–83 (1977).
90. P.H. Rubery, 'Hydrogen ion dependence of carrier-mediated auxin uptake by suspension-cultured crown gall cells', *Planta*, **142**, 203–6 (1978).
91. E. Marre, 'Effects of fusicoccin and hormones on plant cell membrane activities: Observations and hypotheses', in *Regulation of Membrane Activities in Plants*, ed. by E. Marre and O. Ciferri, North Holland Publishing Company, Amsterdam (1977), pp. 185–202.
92. F.A. Smith and J.A. Raven, 'Intracellular pH and its regulation', Ann. Rev. Plant Physiol., **30**, 289–311 (1979).
93. P.H. Rubery and A.R. Sheldrake, 'Carrier-mediated auxin transport', *Planta*, **118**, 101–21 (1974).
94. P.H. Rubery, 'The effects of 2,4-dinitrophenol and chemical modifying reagents on auxin transport by suspension-cultured crown gall cells, *Planta*, **144**, 173–8 (1979).
95. P.J. Davies and P.H. Rubery, 'Components of auxin transport in stem segments of *Pisum sativum* L.', *Planta*, **142**, 211–19 (1978).
96. R.M. Devlin, 'Influence of plant growth regulators on the uptake of Naptalam by *Potomogeton*', *Proc. N.E. Weed Sci. Soc. Philadelphia*, **28**, 99–105 (1974).
97. F. Zsoldos, B. Karvaly, I. Toth and L. Erdei, '2,4-D induced changes in the K^+ uptake of wheat roots at different pH values', *Physiol. Plant*, **44**, 395–9 (1978).
98. J.A. Raven, 'Transport of indoleacetic acid in plant cells in relation to pH and electrical potential gradients, and its significance for polar IAA transport', *New Phytol.*, **74**, 163–72 (1975).
99. A.G.T. Babiker and H.J. Duncan, 'Penetration of bracken fronds by amitrole as influenced by pre-spraying conditions, surfactants, and other additives', *Weed Res.*, **15**, 123–7 (1975).
100. D.J. Turner and M.P.C. Loader, 'Effect of ammonium sulphate and other additives upon the phytotoxicity of glyphosate in *Agropyron repens* L. Beauv.', *Weed Res.*, **20**, 139–46 (1980).
101. L.V. Edgington, 'Structural requirements of systemic fungicides', *Ann. Rev. Phytopath.*, **19**, 107–24 (1981).
102. M.H. Zimmermann and J.A. Milburn, *Transport in Plants. I. Phloem Transport*, Vol. 1, Encyclopedia of Plant Physiology New Series, Springer, New York, Heidelberg, Berlin (1975)
103. U. Lüttge and N. Higinbotham, *Transport in Plants*, Springer-Verlag, New York, Heidelberg, Berlin (1979), 468pp.
104. Jacinta Crowley and G.N. Prendeville, 'Effects of herbicides of different modes of action on leaf-cell membrane permeability in *Phaseolus vulgaris*', *Can. J. Plant Sci.*, **60**, 613–20 (1980).
105. R.C. Kirkwood, 'Action on respiration and intermediary metabolism', in *Herbicides, Physiology, Biochemistry, Ecology*, ed. by L.J. Audus, Vol. 1, Academic Press, London and New York (1976), pp. 443–92.
106. D.E. Moreland, 'Mechanism of action of herbicides', *Ann. Rev. Plant Physiol.*, **31**, 597–638 (1980).
107. P.G. Bartels and C.W. Watson, 'Inhibition of carotenoid synthesis by fluridone and norflurazon', *Weed Sci.*, **26**, 198–203 (1978).
108. P.W. Wightman and C. Haynes, 'The mode of action and basis of selectivity of diflufenican in wheat, barley and selected weed species', *Proceedings of British Crop Protection Conference on Weeds* (1985), pp. 171–178.
109. F.M. Ashton and A.S. Crafts, *Mode of Action of Herbicides*, 2nd ed., Wiley, New York and London (1981), 525pp.
110. P.C. Kearney and D.D. Kaufman, *Herbicides: Chemistry, Degradation and Mode of Action*, Marcel Dekker Inc., New York (1975), 2nd ed., Vol. I, 500pp.
111. P.C. Kearney and D.D. Kaufman, *Herbicides: Chemistry, Degradation and Mode of Action*, Marcel Dekker Inc., New York (1976) 2nd ed., Vol. II, 536pp.
112. R.C. Kirkwood, 'The relationship of metabolism studies to the modes of action of herbicides', *Pestic. Sci.*, **14**, 453–60 (1983).
113. L.M.L. Thompson, G.E. Sanders and K. Pallet, 'Experimental studies into the uptake and translocation of foliage-applied herbicides', *Aspects Appl. Biol.*, **11**, 45–54 (1986).
114. R.H. Bromilow, K. Chamberlain and G.G. Briggs, 'Techniques for studying the uptake and translocation of pesticides in plants. *Aspects Appl. Biol.*, **11**, 29–44 (1986).
115. D. Silcox and P.J. Holloway, 'Epidermal stripping techniques and their application to studies of the foliar penetration of nonionic surfactants', *Aspects Appl. Biol.*, **11**, 19–28 (1986).

116. Irene McKay and R.C. Kirkwood, 'Techniques to establish the effect of foliage-applied herbicides on pigment synthesis and photosynthetic activity of leaf tissues', *Aspects Appl. Biol.*, **11**, 195–203 (1986).
117. D. Coupland, 'Sample preparation for liquid scintillation counting', *Aspects Appl. Biol.*, **11**, 55–66 (1986).

2 Behaviour of insecticide deposits and their transfer from plant to insect surfaces

Martyn G. Ford and David W. Salt
Portsmouth Polytechnic, Portsmouth

2.1 Introduction

The behaviour of insecticides applied to plants grown under field conditions is both complex and variable[1]. The processes that modify the action of the toxicant on the plant surface can occur sequentially or simultaneously, usually resulting in a decreased activity. These processes may broadly be classified as (1) transfer processes, whereby insecticide is transported actively or passively from one location to another, and (2) processes of chemical modification, which produce qualitative changes to the structure of the applied insecticide. The effectiveness and selectivity of insecticide treatment will depend upon the rates at which these two sets of processes occur and their interaction. True equilibria are seldom established, because only a limited time is available to protect a crop from pest populations before levels of economic damage are reached.

To understand the way insecticides act in the field, it is necessary to study their dynamic behaviour on plant surfaces; the mechanisms of transfer and chemical modification may then be identified and quantified. Such an understanding should result in a more deductive, less empirical approach to the design of application and formulation methods than has so far been feasible. The last two decades have witnessed a considerable increase in such research, as new methods of use become necessary to exploit the advantage of insecticidal deposits which can control pest populations at application rates of a few nanograms per square centimetre of treated surface. Moreover, the requirements of selectivity, controlled persistence and low environmental contamination have all focused attention on the detailed behaviour of deposits in the field. This review attempts to summarize the results of this work in a critical manner. It is hoped that this account will suggest areas of research deserving of further study.

2.1.1 Types of insecticidal application

The extensive range of formulation types developed for pesticide application include many examples appropriate for insecticide use. Materials used for crop protection are normally applied as liquid droplets or solid particles of varying size and chemical constitution. The nomenclature of the various liquid and solid formulations, based upon particle size range, may be classified according to a scheme suggested by Furmidge[2] and outlined in Table 2.1. Liquid formulations comprise an active ingredient dissolved or suspended in a liquid carrier that acts as a diluent. These formulations include wettable powders (WP: solid particles impregnated with active ingredient and suspended in a suitable carrier, usually water) and emulsifiable concentrates (EC: oil solutions of active ingredient dispersed as fine droplets in a continuous aqueous phase containing surfactant). Suspension concentrates (SC) are analogous to emulsifiable concentrates, but consist of *solid* particles of sufficiently small size to be held in suspension when water is added as a carrier. In all these formulations the carrier is volatile and evaporates during flight and after impaction to leave a concentrated deposit of active ingredient on the plant surface. This deposit will consist of solid particles (WP or SC) or liquid droplets (EC) each containing insecticide. More recently, ultra-low volume (ULV) formulations have been developed. These

Table 2.1 Classification of pesticide formulations according to particle size[2]

Formulation	Particle diameter in flight (μm)	Particle diameter in suspension (μm)
Liquid		
Wettable powders	80–500	2–7[a]
Emulsifiable concentrates	80–500	0.2–0.3[a]
ULV formulations	30–120	NA
Suspension concentrates	30–500	≤30[a]
Solid		
Fine dust	< 44	NA
Coarse dust	44–100	NA
Microgranules	100–300	NA
Granules	> 300	NA

[a]Volume median diameter (v.m.d.).
NA — not applicable.

comprise simple solutions of insecticide(s) in non-volatile organic solvent(s) which produce small spray droplets stable during flight. The droplets remain as liquid deposits after impaction. Adjuvants may sometimes be incorporated to improve performance (e.g. acrylamide polymers to reduce drift, light-absorbing materials to reduce photodegradation, volatile low molecular weight solvents to reduce viscosity during drop formation). Solid formulations contain the active ingredient mixed with an inert carrier that acts as a diluent and may also contain adjuvants such as surfactants to modify surface tension. A granule or dust is a solid formulation of stated size (Table 2.1) from which the toxicant is transferred to the target organism. Baits, in contrast, retain the toxicant and the pest must encounter and digest the formulation to receive a lethal dose. Regardless of formulation, once an insecticide has been released it is subject to the processes of absorption, decomposition, volatilization and leaching.

2.1.2 Deposit size and transfer

Most insecticides are formulated as liquids and applied as sprays using hydraulic pressure nozzles or atomizers. Spray particles may be electrostatically charged[3, 4] but are generally neutral and fall to the target surface under the influence of gravity, a process termed sedimentation[5]. This method of transfer has traditionally made use of relatively large drop sizes (greater than 200 μm diameter). Large drops will experience a gravitational force sufficiently large to dominate the competing forces of turbulent air movement that occur in the field[1, 6]. Large dimensions, however, while assisting sedimentation, may be less appropriate for impaction and retention, and may give rise to deposits inappropriate for efficient transfer to the target insect. A body volume range of

2×10^7 exists within the class Insecta[1]. Crop pests include insect species that have a range of larval size from less than 1 mg on hatching from the egg to 1 g at the final instar. The relative importance of the forces of gravity, inter-facial tension, internal cohesion and adhesion which determine the retention and accumulation of insecticide by the insect will change with increasing size, as gravitational effects assume a more dominant role. The absolute dimensions both of the deposit and the target insect will therefore determine which forces dominate insecticide transfer.

2.1.3 Deposit behaviour on intermediate surfaces

Although direct interception of sprays by larvae can occur when insecticides are used as eradicants, most agricultural insecticides are applied as protectants to the plant surface, usually the foliage or stem. Transfer and loss can occur from this intermediate surface. In order to understand the range of processes that influence the efficacy of insecticide deposits on such surfaces, the transfer of insecticide from the spray head to the target insect can be represented as a succession of stages, viz. sedimentation, impaction and retention on the plant surface; subsequent availability at the intermediate plant surface; transfer of material from the intermediate surface to the target insect. Each stage has been the subject of investigation, details of which have been reviewed by Hartley and Graham-Bryce[1]: other texts describe different aspects of the subject[7–9]. Most reviews have emphasized the behaviour of insecticide sprays prior to their interception by the plant surface; this review attempts to present the general principles that have emerged from studies describing the behaviour of insecticide deposits and their transfer to insects once impacted on a crop. The evidence suggests that events which occur on the plant surface have a crucial effect upon biological activity.

2.2 Factors determining availability of insecticide at the plant surface

Application of insecticides can result in deposits which vary in both physical state and degree of cover. The physical state of a deposit will depend on many properties, including the physicochemical natures of the active ingredient, plant surface, and insecticide carrier. Although the properties of the active ingredient and those of the plant surface are fixed, the properties of the formulation can be modified to obtain deposits of maximum effectiveness in any particular situation. Formulation therefore provides an important link between properties appropriate for efficient application, and maximum biological efficacy. If the full potential of an insecticide is to be realised, the mechanisms which result in increased availability of insecticide to the target insect and the manner in which these mechanisms change with time, must be understood. The development of more efficient insecticide formulations should then be possible.

2.2.1 Behaviour of deposits at impaction

The behaviour of a deposit on a plant surface will depend on the manner of its impaction on that surface. Airborne drops moving towards a plant have kinetic energies proportional to their masses. On impaction, this energy will be dissipated as damped oscillations of droplet spread and recoil. If, on recoil, sufficient energy remains to overcome the force of adhesion to the surface, the drop will bounce and may subsequently shatter. The extent of oscillation is determined largely by the kinetic energy of the drop and its surface tension, which act together to determine the equilibrium position of the drop. Once oscillations have ceased, liquid deposits will tend to spread across the surface at an initial rate determined by the drop viscosity[10, 11] and the extent to which the underlying surface is wetted. Gravity will influence the rate of spread of large drops ($> 200 \, \mu m$ diameter) but will have less influence on small drops; below $80 \, \mu m$ diameter gravity may generally be ignored. Further complexity is introduced by the dynamic changes in surface tension which accompany the loss of volatile carrier as the deposit dries[1]. All of these processes can modify the physical nature of the deposit. Thus, distribution of active ingredient within the deposit is modified by the volatility of the carrier, and quick-drying deposits may be more uniform than those for which drying is protracted[1]. The surface tension effects observed during drying include droplets running up or down inclined surfaces, lateral movement on horizontal surfaces, annulus formation with associated recoil of the concentrating drop to its centre of spread, and extended annuli which fragment to form series of small droplets[1]. Surface tension changes of as little as $0.02 \, mN/m$ may alter surface energy by as much as the total available gravitational energy. Surface tension can therefore be a dominant force determining the behaviour and properties of liquid deposits.

2.2.2 Wettability of leaf surfaces

The affinity of a deposit for the plant surface will have an important influence on the behaviour of the deposit after impaction and on its subsequent transfer to a target insect. This affinity will determine the extent of wetting of the leaf surface by the deposit, a process related to surface tensions and hence forces of adhesion. Surface tension is commonly investigated by measuring the angle of contact between the surface of a drop and that of the plant. Holloway[12] has presented a detailed account of the wettability of a variety of plant surfaces based on contact angle measurements of water drops placed on these surfaces. Water, which is commonly used as a diluent in insecticide formulations, is a particularly useful liquid for such a study, since it has both a very large dipole character and an ability to act as a donor or acceptor in hydrogen bond formation. Both properties will lead to high adhesion (and hence low contact angles) on polar surfaces with complementary hydrogen bonding properties, but poor adhesion on hydrophobic surfaces with no hydrogen bonding potential. Holloway[12] observed that the external surfaces of plants range from strongly water repellent to completely wettable (Table 2.2). This and other studies suggest that the principle factors that govern the wettability of plant surfaces are the degrees of hydrophobicity[12–14] and the roughness[12, 15] of the surface.

(a) Influence of the hydrophobicity of the plant surface. The hydrophobicity of plant cuticular waxes (Table 2.2) will depend on their chemical composition and on the orientation of their molecules in the solid state[16–18]. Although no class of plant wax is outstandingly water repellent, alkanes (containing only the elements carbon and hydrogen) form the largest contact angles with water (106–109°)[12]. Introduction of oxygen into the hydrocarbon skeleton increases the number of functional groups and introduces dipoles of varying strength. These functional groups prevent close packing and introduce hydrogen bond potential, thus lowering the contact angle formed by the cuticle surface with water droplets (Table 2.2a). Highest contact angles are achieved with the normal alkanes, which in *Brassica oleracea* form the most abundant chemical class present on this plant surface. In general, the contact angle of water falls as the number and nature of the functional groups increase and reaches a minimum (70°) with the normal α, ω-diols (C_{22}–C_{26}). This trend suggests that *Allium porrum* with an abundance of n-alkyl ketones, should produce marginally lower contact angles with water drops than *Brassica oleracea*. Holloway[12] has classified leaf surfaces into two groups, those with contact angles greater and less than 90°. The upper and lower surfaces of the same leaf may belong to different groups, cf. *Trifolium repens* and *Acer pseudoplatanus* (Table 2.2), with the upper surface usually the more hydrophobic.

The behaviour on plant surfaces of oil- and water-based pesticide deposits was studied by Baker *et al.*[19]. They investigated the spread of droplets of known size (50–400 μm diameter) on leaves of major crop species whose adaxial surfaces varied from glaucous to glossy (Table 2.3). A microapplicator was used to produce deposits of various formulations on adaxial surfaces (Table 2.4). On impaction the droplets spread by a factor calculated from the ratio r_2^2/r_1^2, where r_1 is the radius of the in-flight drop and r_2 the radius of the dried deposit. The impact behaviour of volatile aqueous solutions varied with the surface properties of the leaf. Retained droplets (in-flight diameter 175 μm; impaction velocity 7 ms^{-1}) spread readily across the sparsely wax covered, smooth surfaces of dwarf bean, sugar beet and lemon to give spread factors in the range 3.2–4.0. Rapid drying led to deposits for which no further lateral spread occurred. On the hydrophobic surfaces of carnation, clover, strawberry and maize, aqueous droplets had a high contact angle with the leaf surface, which resulted in poor adhesion and a low spread of 1.0–2.1. A droplet with the above in-flight diameter (175 μm) would be expected to spread under the influence of gravity[1]. A spread factor of 1, however, implies no such gravitational effect but suggests

Table 2.2 The wettability of plant surfaces[12]

(a) The effect of wax composition on the contact angle of water droplets

Chemical class	Formula	Homologue range	Contact angle of water drops on single homologue examples of wax constituent (deg)	Composition on two species of plants (%)	
				Brassica oleracea	*Allium porrum*
n-Alkanes	$CH_3—(CH_2)n—CH_3$	Odd C_{20}–C_{35}	106–109	32.3	19.8
n-Alkyl esters	$CH_3—(CH_2)_n—CO—O—(CH_2)_m—CH_3$	Even C_{32}–C_{70}	104–106	9.8	27.2
n-Alkyl ketones	$CH_3—(CH_2)_n—CO—(CH_2)_m—CH_3$	Odd C_{25}–C_{35}	104–106	15.9	30.7
n-Secondary alcohols	$CH_3—(CH_2)_n—CHOH—(CH_2)_m—CH_3$	Odd C_{21}–C_{33}	103–104	13.9	
n-Fatty acids	$CH_3—(CH_2)_n—COOH$	Even C_{14}–C_{36}	101–102		
n-Primary alcohols	$CH_3—(CH_2)_n—CH_2OH$	Even C_{22}–C_{32}	94–95.5		
n-ω-Hydroxyacids	$HOCH_2—(CH_2)_n—COOH$	Even C_{10}–C_{30}	90–95		
Triterpenoids	Various	C_{30}	89–95.5		
Stenols	Various	> C_{24}	82.5–92		
n-α,ω-Diols	$HOCH_2—(CH_2)_n—CH_2OH$	Even C_{20}–C_{32}	70–71		
				Total 71.9	77.7

(b) Contact angles of water on some leaf surfaces before and after chloroform washing, and on smooth films of their isolated superficial waxes. (The contact angle standard deviations are given in parentheses: U = upper, L = lower leaf surface)

Species	Surface	Contact angle	Effect of $CHCl_3$ washing (%)	Contact angle of isolated wax
Narcissus pseudonarcissus	U + L	142° 54' (110')	Reduction 37.6	105° 58' (17')
Eucalyptus globulus	U + L	170° (123')	Reduction 45.6	105° 15' (23')
Trifolium repens	U	158° 52' (76')	Reduction 11.9	102° 27' (26')
	L	10° 57' (62')	Reduction 10.9	
Eucalyptus botryoides	U + L	168° 23' (127')	U reduction 8.3	102° 44' (44')
			L reduction 25.1	
Clarkia elegans	U	124° 8' (276')	Reduction 22.4	102° 35' (30')
	L	159° 15' (85')	Reduction 50.4	

Table 2.2 contd.

Species	Surface	θ1		θ2	
Acer pseudoplatanus	U	44° 30′ (121′)	Increase 123.2	99° 35′ (27′)	
	L	155° 20′ (97′)	Reduction 16.1	103° 15′ (19′)	
Prunus laurocerasus	U	90° 50′ (76′)	Reduction 55.7	103° 30′ (34′)	
	L	93° 32′ (75′)	No change		
Rhododendron ponticum	U	70° 22′ (89′)	Reduction 17.0	104° 58′ (43′)	
	L	43° 21′ (163′)	Increase 41.3		
Saponaria officinalis	U	100° 6′ (47′)	Reduction 34.3	100° 16′ (28′)	
	L	106° 26′ (60′)	Reduction 37.6		
Senecio squalidus	U	90° 10′ (54′)	Reduction 12.8	102° 14′ (42′)	
	L	92° 15′ (61′)	Reduction 14.7		
Rumex obtusifolius	U	39° (121′)	Increase 50.0	100° 45′ (25′)	
	L	40° 5′ (132′)	Reduction 75.0		
Plantago lanceolata	U	74° 23′ (99′)	No change	102° 20′ (82′)	
	L	39° 32′ (241′)	Increase 65.8		

(c) Calculation of the fractions of solid–liquid (f_1) and air–liquid (f_2) interfaces for a number of composite leaf surfaces. (U = upper, L = lower leaf surface)

Species	Surface	θ_1	θ_2	f_1	f_2
Eucalyptus globulus	U + L	105° 15′	170°	0.02	0.98
Tropaeolum majus	L	106° 21′	160° 1′	0.08	0.92
Pisum sativum	U + L	106° 4′	158° 18′	0.10	0.90
Brassica oleracea	U + L	103° 51′	151° 3′	0.16	0.84
Iris germanica	U + L	103° 12′	145° 26′	0.24	0.76
Clarkia elegans	L	102° 35′	124° 8′	0.55	0.45

Table 2.3 Characteristics of the adaxial leaf surfaces of selected plant species[19]

Plant	Surface type	Wax deposit ($\mu g/cm^2$)	Surface morphology
Rape	Glaucous	30–50	Crystalline tubes, plates and dendrites
Pea, clover carnation	Glaucous	10–25	Dense arrangement of crystalline plates
Maize	Semiglaucous	10–15	Crystalline plates
Strawberry	Semiglaucous	25–40	Thick film overlaid by fine wax ribbons
Lemon	Glossy	5–15	Thin film, occasional granules
Sugar beet	Glossy	4– 8	Thin film, occasional granules
Dwarf bean	Amorphous	1– 2	Extremely thin film

Table 2.4 Droplet spread following application of an aqueous solution (0.3%) of 'Uvitex 2B' with in-flight diameter of 175 μm[19]

Plant	Impaction diameter (μm)	Deposit area ($\mu m^2 \times 10^4$)	Spread factor
Rape	56	0.25	0.1
Clover	182	2.6	1.1
Carnation	175	2.4	1.0
Maize	217	3.7	1.5
Strawberry	257	5.0	2.1
Lemon	350	9.6	4.0
Sugar beet	315	7.8	3.2
Dwarf bean	336	8.8	3.7

instead that surface tension is dominating the behaviour of the droplet to result in a profile with minimum contact with the leaf surface. Thus, the effect of gravity on droplet spread will itself be influenced by the affinity of the drop for the leaf surface. Drops with high affinity (and low surface tension) should show a significant gravitational effect, even for small drop diameters, whereas drops with low affinity (and high surface tension) should only be weakly influenced by gravity, even at quite large drop diameters.

The spread of polar drops is opposed by non-polar plant surfaces but facilitated by more polar wax-free surfaces. This relationship is reversed for non-polar oil deposits. The spread of similar size oil droplets, for example, was restricted on the smooth, relatively wax-free surface of sugar beet and lemon leaves (spread factor 1.0–1.2) but was more rapid across the waxy leaf surfaces of rape (spread factor 2.3), strawberry (spread factor 3.7) and maize (spread factor 8.0). On smooth leaf surfaces relatively free of wax, spread was complete within seconds of impaction, whereas on glaucous or hairy leaves the movement of oil droplets was prolonged by the thick wax layer and variable leaf topography. The redistribution of oil droplets on maize, for example, continued for up to 6 hours to result in a deposit area five times greater than the initial area of impaction[19]. Droplets comprising permethrin dissolved in oil move over and into the surface of waxy cabbage leaves during a 10-minute period immediately after impaction, whereas the initial droplet spread on less waxy cabbage leaves appeared complete within seconds of impaction[20] and resulted in stable drop profiles proud of the plant surface. High values of spread factor between 500 and 2500 were occasionally observed for oil droplets applied to maize leaves and attributed to surface differences between mature and immature leaves[19]. Spread factors decreased twenty-fivefold to thirtyfold as senesence approached, presumably because epicuticular waxes or surface roughness were modified.

The use of wetting agents to overcome surface tension barriers between leaf and deposit usually results in a more intimate contact with cuticular waxes. If an intact layer of insecticide in intimate contact with wax-covered leaves is required, pesticide sprays should have a surface tension of about 30 mN/M. This value is obtained for oil drops such as occur in EC and ULV formulations, but for WPs wetting agents are required in the formulation if this value is to be realized. The dynamic tension of liquid deposits, however, is critical in determining the initial behaviour of the impacted deposit, particularly during drying. Application of emulsifiable concentrates to waxy leaf surfaces such as rape can result in the formation of deposits embedded in the epicuticular wax. The structure of the epicuticular waxes is often modified at the perimeter of the deposit to form an annulus of softened wax consisting of a thick amorphous crust with occasional mounds. Most of the active ingredient, if hydrophobic, will be associated with the annulus[19]. In contrast, the wax at the centre of such a deposit may differ little from that of the untreated leaf surface. On polar leaf surfaces, different behaviour is observed. Drops of EC spread rapidly over the relatively non-waxy sugar beet leaf surface, for example, to form a thin continuous cover. Drying forces operate soon after impaction to result in distribution of polar molecules preferentially to the central deposit, whereas non-polar molecules are held in the outer zone in association with the organic solvent. Baker et al.[19] suggest that the redistribution of active ingredient within the deposit depends upon partition between the various phases during droplet drying. Radiolabelled DDT, for example, was located predominantly in the outer annulus with other lipophilic constituents.

The amount of wax present on the surface is also important in determining wettability. The rate of spread of drops (80 μm diameter) of refined mineral oil (Sirius M40, M100 and M350) on cabbage leaf surfaces of different wax cover has been studied by Crease et al.[10]. Drops of low viscosity spread on impaction until the oil had penetrated the wax cover. Counts of the number of drops retained proud of the wax surface two minutes after impaction showed that there was a critical wax cover (approximately $30 \mu g/cm^2$) below which all of the drops persisted, but above which all penetrated into the underlying wax. The extent of spread, however, was substantially less for high viscosity oils (e.g. Sirius M350), even on surfaces of high wax cover.

(b) Influence of the organization and distribution of surface waxes. Removing the superfical wax by chloroform extraction normally reduces the contact angle of the depleted surface, confirming the role of the wax (Table 2.2b). However, the isolated waxes recovered by chloroform extraction of leaf surfaces with a wide range of contact angles (39–170°) tend to a constant angle in the narrow range 99–105° when deposited onto flat glass surfaces, suggesting that it is the organization of the waxes as well as their composition that may determine surface wettability[12].

An example of the way in which the organization of waxes can influence the behaviour of liquid deposits is provided by the studies of Crease *et al.*[10]. These authors observed that the spread with time of Sirius oil droplets on cabbage leaf surfaces could be described by either single exponential equations, for leaves of low wax cover, or double exponential equations, for the more waxy leaves. This suggests that a proportion of the drops landing on waxy leaves impact on an area with surface properties that impart very low droplet stability. The remainder alight on areas with different attributes, where droplets retain large contact angles and stand proud of the surface for long periods of time. Even on very waxy leaves coherent cover may not be achieved (Figure 2.1). Waxy surfaces are therefore characterized by local variation in surface properties. Crease *et al*[10] suggest that this variation is due to the presence of incoherent regions of wax bloom which are more frequent on leaves of high wax cover. The importance of bloom in reducing the persistence of oil drops was underlined by the survival of drops landing on an area of a leaf where the bloom had been destroyed without the removal of the surface wax. Under these conditions, Sirius oil drops landing on the untreated bloomed surface disappeared instantly, whereas those landing on the treated surface of the same leaf persisted for several days, even though the extent of the leaf wax cover was the same.

(c) Effect of surface topography. The wettability of leaf surfaces therefore reflects both the amount and the chemical composition of the superficial wax layer (Table 2.2a). However, neither can account for the observed range of plant cuticle wettability [12], and the roughness or surface topography must also be important. Evidence to support this view is provided by the extreme values of contact angel (less than 70° or greater than 110°) observed for plant surfaces characterized by surface roughness.

The topography of a plant surface modifies the spread of retained droplets by changing the effective area of contact between the deposit and the underlying surface. Topography or roughness may result in (1) composite surfaces comprising solid–liquid and air–liquid interfaces when wetted or (2) non-composite surfaces comprising only solid–liquid interfaces. Surface roughness can either

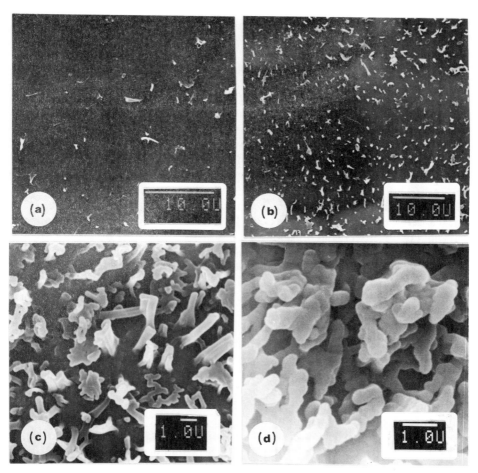

Figure 2.1. Scanning electron micrographs of the surface of cabbage leaves (*Brassica oleracea*) showing the relationship between the density of wax cover ($\mu g/cm^2$) and the proportion of the surface covered by crystalline wax deposits. (a) Wax density 2.5 $\mu g/cm^2$, proportion cover 0.033. (b) Wax density 5.0 $\mu g/cm^2$, proportion cover 0.070. (c) Wax density 18.3 $\mu g/cm^2$, proportion cover 0.624. (d) Wax density > 30 $\mu g/cm^2$, proportion cover 0.956. Dimensions are indicated by solid bars ($u = \mu m$).

increase or decrease the area of contact depending on the conditions. Increasing roughness tends to increase contact angles greater than 90° but reduces angles less than 90°[1]. The contact angle for water drops resting on surfaces of paraffin wax increases from 110 to 158° as the surface becomes more irregular[18]. The relationship between roughness and contact angle for non-composite surfaces may be summarized by the Wenzel equation[15],

$$r = \cos(\theta_2)/\cos(\theta_1)$$

where θ_1 is the actual and θ_2 the apparent contact angle, and for composite surfaces by the equation of Cassie and Baxter[21],

$$\cos \theta_2 = f_1 \cos \theta_1 - f_2$$

where the actual and apparent contact angles are expressed in terms of solid–liquid and air–liquid surface areas (f_1) and f_2 respectively (Table 2.2c). This last relationship takes into account the reduced area of contact between a repelled drop and a rough surface as the result of air trapped at their interface. Microscopic roughness due to leaf venation or epidermal cell boundaries can produce composite surfaces that are water repellent, even after chloroform washing (Table 2.2b). Trichomes, which show large variation in size, shape and frequency per unit area, play an important role in determining the composite nature of leaf surfaces and hence their wetting properties. Closed patterns of trichomes produce composite surfaces that trap air between the drop and the plant surface to result in water repellency[22], whereas open patterns enhance wetting, possibly as a result of capillarity. Non-composite open surfaces tend to be most easily wetted and the layers and crusts associated with such surfaces tend to result in low contact angles (between 80 and 100°)[23]. Superficial deposits of wax can result in surface blooms that produce large contact angles for water drops (greater than 120°)[23–25]. Hall *et al.*[23] report that leaf wax from the genus *Eucalyptus* recrystallized from chloroform extracts of leaf surfaces gave deposits on glass with structures similar to those observed on the original plant surfaces by electron microscopy. Gukel and Synnatschke[26] cite eighteen publications that report contact angle measurements on different plant surfaces. Although probably not equilibrium contact angles, these values provide information about the wettability of a variety of leaf surfaces. They conclude that leaves have surface structures (e.g. wax crystals and blooms, cell margins, hairs, veins and stomata) that depend on the foliar type and age. These are the principle topographic factors to be considered if an overall wettability value for a particular leaf type is to be obtained. In contrast to the findings of Holloway[12], this study suggests little difference between the wettability of the abaxial and adaxial surfaces of some leaves[26]. Closely related plants, such as brassicas or cereal species, wet in a similar manner.

(d) Use of microscopy to study the influence of wetting on deposit availability. Many of the processes described in the preceding sections have been observed on treated plant surfaces when examined under the microscope. Such studies demonstrate the variety of effects that can occur under different conditions. Hart[27], for example, has described the use of the scanning electron microscope (SEM) to investigate the ways in which surface topography may influence the behaviour of insecticides applied as small droplets in liquid or solid carriers. His results are in broad agreement with theoretical considerations. A complex range of surface roughness was reported (Figures 2.2 to 2.5) which varied from smooth surfaces, disrupted only by shallow grooves produced at cell bondaries (e.g. the

adaxial surface of lucerne, *Medicago sativa*), to surfaces containing tapered trichomes at a density of 55 mm^{-2} (e.g. the abaxial surface of melon, *Cucumis melo*). The size of trichomes on melon leaves range from 100 µm to over 1 mm in length, and with a base diameter of approximately 50 µm. An additional form of roughness, small papillate structures 3 µm in height with 1 µm basal diameter at an approximate density of 40×10^3 mm^{-2}, was observed on the adaxial surface of rice (*Oryza sativa*). In nearly all the examples described by Hart, surface roughness was characterized by the shallow grooves (cell margins) that lie between adjacent cells.

Surface roughness is an important determinant of spray retention; it is responsible for droplet bounce by presenting surfaces inclined from the horizontal, and governs the magnitude of contact angle hysteresis. Roughness can also cause droplets to coalesce to form larger deposits. Droplets behaved in this manner when landing on abraded regions of the adaxial surface of horizontal barley leaves (Figure 2.2). On rye grass (*Lolium perenne*), deposits appeared to be retained at the top of the longitudinal grooves, leaving the valley bottoms relatively free of pesticide. This behaviour suggests that water-based droplets form with an air cavity isolating the valley bottom from the formulation: only the groove apices are then wetted. In contrast, oil-based droplets deposited on the hairy but relatively polar adaxial surface of cotton retain their integrity and, as a result of a large contact angle, remain proud of the leaf surface, where they are available to a moving insect (Figure 2.3a).

Surface roughness can also influence the final distribution of a deposit. The organic phase of insecticide formulations is often located preferentially along the cell margins with a high concentration held within a thin annulus at the deposit perimeter. Formulations based on solid particles leave deposits that lie superficially on the plant surface and are less intimately associated with the epicuticle. Retention then depends upon rather weak forces of adhesion. Adhesion of wettable powder formulations to leaf surfaces, for example, is poor for many species: on drying, such deposits tend to form a crust of particles lying loosely upon the wax layer with the active ingredient more tightly bound to the inert support material[19].

Surface hydrophobicity determines the strength of adhesion of deposits to competing surfaces. Thus, oil-based formulations of pirimicarb sprayed on the polar abaxial surface of broad bean leaves (*Vicia faba*) (Figure 2.3c) do not spread, but remain available above the leaf surface. Contact of a drop by the black bean aphid (*Aphis fabae*) resulted in a thin film being transferred to the insect which spread, pulling the legs and antenna into close proximity to the body. This behaviour indicates a high adhesion force between the oil and the surface of the aphid. The same oil-based formulation of pirimicarb, when applied to the adaxial surface of a young cabbage leaf, spread to wet the surface waxes almost completely (Figure 2.3b). There was some evidence of similar behaviour in a small percentage of drops deposited on the abaxial surface of the broad bean leaf, although the overall probability of spreading appeared low. Dried drops of a

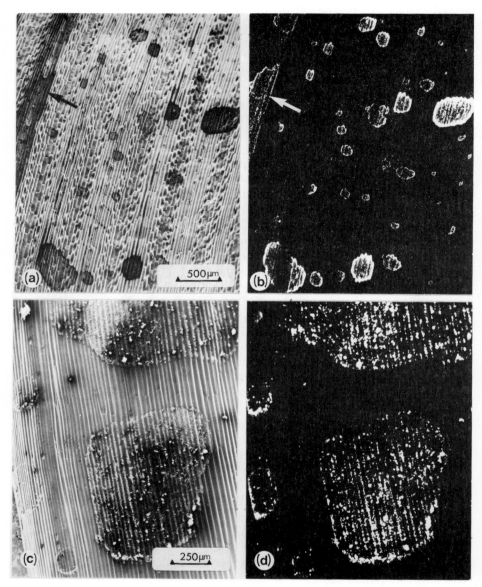

Figure 2.2. Monocotyledon leaf surfaces showing the nature of pesticide deposits applied as conventional high-volume aqueous sprays. (a) Secondary image and (b) cathodoluminescent image of the adaxial surface of barley leaf (*Hordeum vulgare*) sprayed with an ethirimol formulation of the fungicide Milgo (300 l/ha; 0.15 per cent concentration). Droplets adhering to regions of abraded wax can coalesce to form larger deposits (arrows). (c) Secondary image and (d) cathodoluminescent image of the abaxial surface of rye grass leaf (*Lolium perenne*) sprayed with a dispersion of an experimental growth regulator.

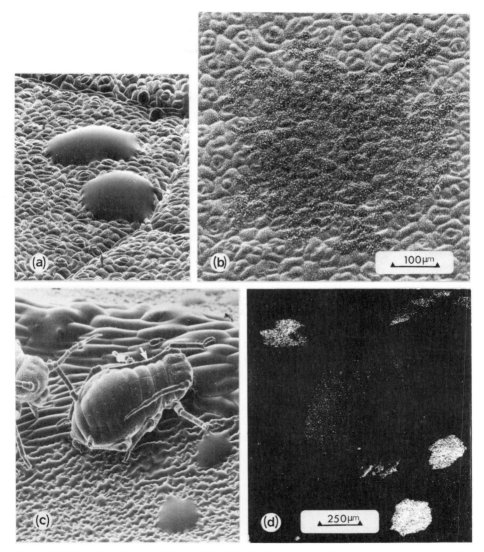

Figure 2.3. Scanning electron micrographs of leaf surfaces sprayed at ULV with the insecticide pirimicarb dissolved in a low-volatility carrier oil. (a) ULV droplets on the relatively non-waxy adaxial surface of cotton leaf (*Gossypium hirsutum*). (b) A similar drop at the same magnification spread out on the waxy adaxial surface of a cabbage leaf (*Brassica oleracea*). (c) Secondary image and (d) cathodoluminescent image of a similar formulation of pirimicarb sprayed on the abaxial surface of the broad bean leaf (*Vicia faba*) on which black bean aphid (*Aphis fabae*) was feeding. Drops adhering to the leaf cuticle can be observed as discrete deposits whereas those contacted by the insect spread to a thin film covering the insect surface and causing antennae and legs to adhere to the thorax and abdomen (arrow). The contrast between discrete and spread deposits is most clearly observed in (d).

non-ionic surfactant applied to the adaxial surface of wheat leaves (*Triticum aestivum*) (Figure 2.4) were deposited on the epicuticular wax to give a low EM secondary electron signal after 1 hour. Similar deposits at 7 and 48 hours respectively gave a larger secondary electron signal consistent with penetration of surfactant into the epicuticular waxes to leave the rougher more reflective leaf surface exposed to the beam of electrons.

The spread of drops on plant surfaces can lead to an increased encounter probability as the deposit area increases but a reduced likelihood of the deposit being transferred to an insect as the active ingredient becomes more intimately associated with the underlying surface. Surface tension presents the main barrier to droplet spread on surfaces with a low affinity for the deposit. Wetting agents can effect the spreading of deposits, but their use may have unforeseen and unwanted consequences. When single drops of solutions of a surface active agent were dried on leaf surfaces, the degree of spread was observed to depend on both the leaf type and the growing conditions. Micrographs of pesticide deposits taken several hours after spraying suggested that, at this time, much of the surfactant had penetrated into the leaf, leaving a reduced surface deposit (Figure 2.5). Aged deposits of this type are difficult to identify and will be less available to insects. Addition of a fine dispersion of a cathodoluminescent phosphor to the carrier allowed measurement of the original distribution of the drop. When the water-soluble salts of organic acids are sprayed onto the relatively non-waxy abaxial surfaces of barley (*Hordeum vulgare*) and ryegrass (*Lolium perenne*), the drops spread sufficiently to coalesce, indicating a high adhesive force between the droplet and leaf surface (Figure 2.2). This result suggests that the cuticle beneath the wax is formed of relatively polar material. The presence of epicuticular wax, however, will change the surface layer to a more hydrophobic character. Thus, oil-based droplets of the aphicide pirimicarb spread to a thin film on the waxy surface of a cabbage leaf (*Brassica oleracea*) and on the waxy cuticle of a black bean aphid (*Aphis fabae*) but remained as discrete drops proud of the surface when placed on the relatively non-waxy leaves of cotton (*Gossypium hirsutum*) and broad bean (*Vicia faba*) (Figure 2.3). Transfer is only adversely affected if droplet spread results in the deposit becoming too intimately mixed with the constituents of the plant surface.

2.2.3 Penetration into the plant cuticle

The availability to target insects of contact insecticides on plant surfaces will depend upon the deposits remaining at the surface: accumulation of insecticide by a target insect will be reduced by transfer of material into the plant. For good contact action, therefore, insecticides should be deposited in a manner that restricts their foliar penetration. Foliar penetration is considered in some detail in the accompanying review by Kirkwood (Chapter 1), and although the work described in his article concerns herbicide penetration, the general principles he describes are also appropriate for the behaviour of insecticide deposits. Penetration of organic materials is determined by their physicochemical properties and those of

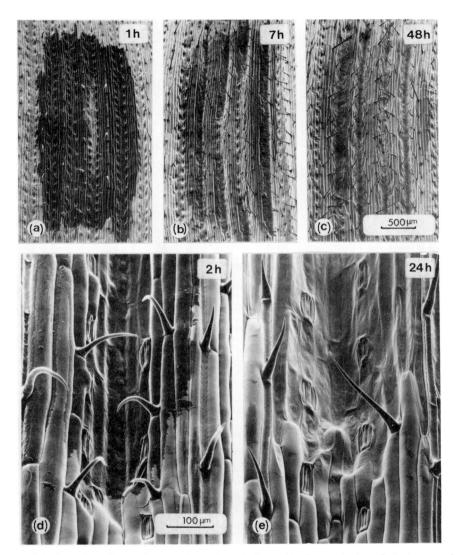

Figure 2.4. Secondary electron images of aged deposits formed by the non-ionic surfactant syperonic NX (1 μl of 0.2 per cent m/m) applied to the adaxial surface of wheat leaves (*Triticum aestivum*) subsequently held at constant relative humidity (85 per cent) and temperature (15–17 °C). (a) One hour after application the deposit appeared as a dark region resting on the epicuticular wax. An area in the deposit centre appeared devoid of surfactant. As the deposit aged, (b) and (c), the dark region beame lighter as the surfactant was taken up into the leaf. (d) Soon after application (2 h) the epidermal cells and stomata had collapsed as a result of surfactant uptake. (e) By 24 h the epidermal cells above the veins had also collapsed. Very little surfactant was present on the leaf surface by this time.

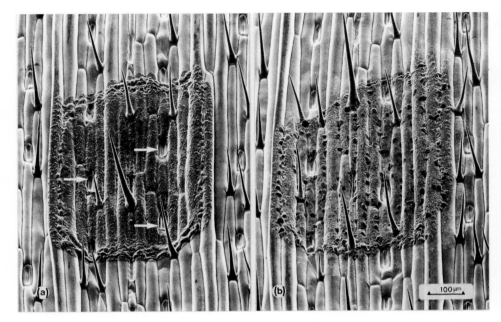

Figure 2.5. Secondary images of aged deposits formed by a diluted suspension concentrate of the fungicide Milgo (0.2 μl drop) applied to the abaxial surface of a wheat leaf (*Triticum aestivum*). During ageing, the relative humidity was 85 per cent and the temperature was 16–20 °C. (a) A deposit observed 30 min after application: the stomata remain unwetted (arrows). The relatively polar active ingredient was deposited on the surface with other formulation ingredients including the waxy non-ionic surfactant Cirrasol ALN-WF. Note the thicker deposit at the perimeter suggesting surface tension effects during drying. (b) A similar deposit aged five days. The density of the deposit was reduced, presumably by surface uptake. Particulate ethirimol (the active ingredient) remains proud of the surface, suggesting that the more waxy formulation constituents may have penetrated at a faster rate.

the plant surface. The important characteristics are briefly summarized below: however the reader is referrred to Kirkwood's text for a more comprehensive review than has been attempted here.

1. The cuticular wax layer appears to be the main barrier to the penetration of polar charged materials[28, 29], whereas the more polar material underlying this layer restricts the movement of non-polar materials[1].

2. The capacity of the cuticular wax layer for non-electrolytes leads to their slow dissolution in this layer, where they are retained[30, 31] close to the leaf surface and where they may then come into contact with a target insect[1]. Dissolution may be a penetration rate-limiting process[32].

3. Because the wax layer is usually a polymorphous solid[13], movement of dissolved material is slow. The rate of penetration in this layer is probably diffusion controlled, with diffusion coefficients lying in the range 10^{-9}–10^{-12} cm²/s. Such values correspond to movement of material over a distance of only a few millimetres in a few days or weeks. These rates are similar to those

Table 2.5 Predicted volatilization at 20 °C of insecticide deposits from 1 ha of inert plane surface

Compound	Molecular weight	Vapour pressure (Pa)	Saturated vapour concentration (g/ml)	Loss in 1 hour
Naled	381	2.7×10^{-1}	4×10^{-8}	5 kg
Diazinon	304	1.9×10^{-2}	2.3×10^{-9}	290 g
Fenthion	278	4.1×10^{-3}	4.6×10^{-10}	58 g
γ-BHC	291	1.3×10^{-3}	1.5×10^{-10}	19 g
Parathion	291	7.8×10^{-4}	9.4×10^{-11}	12 g
Dieldrin	381	3.5×10^{-4}	6.5×10^{-11}	8.2 g
Permethrin	391	2.0×10^{-5}	3.3×10^{-12}	0.4 g
Phosfolan	255	7.8×10^{-7}	7.9×10^{-14}	10 mg
Cypermethrin	416	5.3×10^{-10}	8.8×10^{-17}	11 μg

observed for the lateral movement of non-volatile insecticides away from the initial site of impaction[33].

This behaviour explains why most insecticides act as persistent contact or stomach poisons, and suggests that systemic action will be restricted to a few polar materials that have small octanol/water partition coefficients[34]. Because retention at the plant surface is a crucial feature of the contact toxicity of an insecticide, penetration of materials into the plant is an area of research worthy of further study.

2.2.4 Evaporative loss of volatile materials

Availability at the plant surface may also be reduced by evaporation of the active ingredient. Evaporative loss of insecticide from deposits resting on plant surfaces can be significant, provided the compound is sufficiently volatile. Although vapour pressures at 20 °C must generally be equal to or greater than 10^{-2} Pa for a toxic fumigant action, loss of residues from plant surfaces can occur for materials with vapour pressures lower than this value. In fact, theoretical considerations (Table 2.5) suggest that vapour loss is potentially important for all but the least volatile materials (with vapour pressures at 20 °C less than 1×10^{-5} Pa). The rate of loss will depend upon the surface area of the deposit, regardless of the extent of the deposit cover. The surface area of a deep, uniform deposit, for example, will remain more or less constant until the deposit becomes sufficiently thin to result in incomplete cover of the plant surface. The initial rate of loss can therefore be expressed as

$$-\frac{dM}{dt} = \text{constant} \tag{2.1}$$

Similarly, isolated spots of identical size will also evaporate at a rate proportional to surface area, which will be a function of the deposit radius

$$\frac{dM/A}{dt} = 2naD \tag{2.2}$$

where M is the insecticide mass, A is the area of the deposit, n is the number of spots of radius a and D is the vapour diffusion coefficient. These two forms of deposit represent extremes of cover, which should both be characterized by linear rates of evaporative loss. Such behaviour, however, has seldom been reported[1]. In fact, observation suggests that the process is characterized by a diminishing rate as time passes. This result suggests that loss from plant surfaces is a more complex process than simple volatilization of molecules from the surface of a deposit. The overall rate of evaporation is likely to decrease, for example, as surface material dissolves in the epicuticular wax, and the availability of insecticide for volatilization at the plant–air interface is consequently reduced.

Phillips[35] described the persistence of crystalline deposits of aldrin and dieldrin. Non-linear loss of insecticide observed from glass surfaces could be approximately described by exponential equations of the general form:

$$y = A_1 e^{-k_1 t} + A_2 e^{-k_1 t} \ldots + A_j e^{-k_j 5} \tag{2.3}$$

where y is the quantity of insecticide remaining on the leaf surface at time t $\sum_{i=1}^{i=j} A_i$ is the quantity on the surface at $t = 0$ that is available to decay at a rate determined by the rate constants for loss, k_i. The subscript j denotes the number of exponential terms included in the equation and may represent the number of first-order processes contributing to the overall loss of insecticide from the leaf surface. The rates of observed loss were consistent with theoretical values presented in Table 2.5. The greater the surface area of a deposit the greater its initial rate of volatilization but the greater its probable transfer to the target insect. Hence formulations with spray properties that maximize surface area will maintain a high encounter probability, but for a reduced period of time after application, since evaporation (and also penetration) will be enhanced. Any procedure that reduces volatility should result in a significant increase in the persistence of treatments. Liquid or solid carriers may substantially reduce evaporation by reducing the initial vapour pressure and giving rise to sigmoidal rate curves[35]. The mechanism of this reduced vapour pressure probably involves diffusion control of the insecticide concentration in the stagnant layers at the deposit–air interface and intermolecular attraction between the insecticide and the carrier molecules. Wind speed greatly increases evaporative losses but a limiting value will be reached; increased wind velocity will then have no further effect on the evaporation rate. However, quite low wind speeds (3–5 km/h) can have profound effects on pesticide loss. Surface roughness, which enlarges surface area, increases loss by volatilization in a manner that is enhanced by conditions of high atmospheric humidity: the reason for this is obscure, although Phillips[35] has suggested that the mechanism may involve competition between the insecticide and water molecules for sites on the treated surface.

Evaporative loss can be affected by the properties of the underlying surface. The overall shape of curves describing the loss of insecticide from cotton leaves can be similar to those found for the loss of the material from glass surfaces[36]. However, complete loss from foliage is not attained, even after a long time, because some insecticide is retained at the plant surface by the epicuticular waxes[30], or even in the cutin layer itself[1]. Differences in rates of loss from surfaces of different plant types might be expected since the quantity of surface waxes, which varies between species, can range from less than 1×10^{-6} to 1×10^{-4} g/cm^2[19, 28]. Plant waxes can therefore have considerable capacity to retain insecticides in solution.

The non-linear kinetics of evaporative loss may arise from other mechanisms, however. Hartley and Graham-Bryce[1] have considered the persistence of residues comprising a population of deposits of different size. Although, in such a population, each individual spot would evaporate at a constant initial rate, the rate of loss of the total residue would depend upon the sum of the individual evaporation rates for each spot. By changing the relative frequencies of the drop sizes, these authors generated a set of curves that ranged from an approximately linear form for equal numbers of different drop sizes (a rectilinear distribution) to the commonly observed non-linear form for a simulated symmetric distribution with maximum frequency at the mean drop size. They point out that the progressive disappearance of the smaller deposits from a mixed population will always result in a steeper curve than that for a population of uniform size. These results emphasize the importance of the statistical properties of processes governing insecticide behaviour, particularly under field conditions where variation can be substantial.

2.2.5 Weathering of insecticidal deposits

Loss of insecticide deposits by weathering may be the result of mechanical detachment by wind or rain, leaching by rainfall or chemical degradation as a result of the action of sunlight. In order to dislodge a particle or droplet that has impacted on a plant surface, the detachment force must overcome the force of adhesion to the surface. If this is achieved, the particle will move to a new site where it will rest until again detached. Eventually, the particle will come to rest in a stable position of low potential energy[1].

The resistance of an insecticide deposit to mechanical detachment will therefore be related to its adhesion to the underlying plant surface. Adhesion of solid particles depends largely on the presence of a liquid film which can wet both the particle and the plant surface. Adhesion will remain constant until the liquid has completely evaporated, but will then decrease dramatically. A permanent adhesion will occur, however, if, on drying, a coherent layer of non-volatile liquid is formed between the deposit and the underlying surface. With liquid deposits, of course, good adhesion is maintained for as long as the deposit remains in this state. Because deposits range in size from 20 to 500 μm diameter, they have low mass and only experience a small gravitational force. As a result, the adhesive forces needed to prevent detachment are quite small. Adhesives or stickers can be added to a formulation to achieve more permanent retention. Adhesives are usually low molecular weight polymers which can dissolve in the liquid film but retain their flexibility on drying. The addition of amine stearates to pesticide formulations, for example, is claimed to increase the adhesion and hence the retention of the spray particle[1]. The adhesion of both liquid and solid deposits will depend on the extent of the contact achieved between the deposit and the underlying surface.

Leaching by the action of rainfall can be important in determining persistence. Water-soluble constituents will dissolve in rainwater that has settled on wetted leaf surfaces. Under ideal conditions of coherent water, 1 cm of rain forming a thin film moving across the surface of a wetted leaf surface could dissolve as much as 100 mg/ha of an insecticide with a water solubility of 1 µg/l. Such calculations[1] are based on the assumption that the deposit is completely immersed in the water film. In practice, dry insecticide particles often repel water, which then flows around the deposit, reducing the extent of leaching. ULV deposits formed from oil-based formulations are less likely to be wetted and leached by rainfall than emulsifiable concentrates or wettable powders containing surfactants. A number of reports based on both field and laboratory studies give tentative support to this view. Laboratory studies by Crease et al.[10], for example, have shown that the biological efficacy of Sirius oil deposits of cypermethrin on cabbage leaves, later subjected to light rainwashing, persisted more effectively when the active ingredient was formulated in a viscous oil. This result is probably related to the ability of low viscosity oils to spread rapidly over the surface of water drops, minimizing surface energies at the air–water interface. The reduced rate of spread associated with viscous oils probably limited the extent to which this process could occur. Rainfastness may also depend upon the size of the individual deposits. ULV drops of methyl-parathion, with a drop diameter of 100 µm, for example, were more resistant to rainwashing than those of larger diameter. Deposits containing surfactants may be more extensively leached than oil-based formulations[37] since surfactants should overcome the high surface tensions between non-polar formulations and rainwater[38]. However, surfactants increased association of DDT with cuticular wax layers following rainwashing. Insecticide was presumably dissolved in the surfactant deposits visible on treated leaf surfaces[31].

A more detailed study of the effects of natural and artificial rainwashing on the persistence of various insecticides formulated as emulsifiable concentrates or wettable powders has recently been reported by Pick et al.[37]. Although effects were confounded, a number of tentative conclusions can be drawn. Rainwashing can have a severe effect on persistence, particularly if the deposit is fresh. The effect varies with formulation. For most pesticides approximately half of the original deposit applied one hour earlier can be removed by rainfall of a few minutes. Ageing deposits become more rainfast, probably as material penetrates into the leaf surface. Deposits formed from emulsifiable concentrates mix with the wax layer and show increased rainfastness as they age; wettable powders, on the other hand, stay proud of the surface, remaining susceptible to rainwashing. The intensity of rainwashing (volume per unit time) was unimportant, and loss was related primarily to the absolute volume of rain (measured in millimetres) which fell. The intensity of rain observed in these experiments was high. Intensity may therefore have some effect at lower levels. In contrast to earlier reports[31], the presence of surfactant had no obvious effect on rainfastness. These studies suggest that further investigation of the relationship between rainfastness and the ageing process of insecticide deposits would be desirable.

Weathering can also result in chemical modification or degradation of insecticides with loss of biological activity. Photocatalysed reactions are probably responsible for most degradation processes that occur at the plant surface. Any chemical whose maximum absorbances lies in the visible region of the spectrum is potentially photodegradable. Although sunlight has little energy below 290 nm (as a result of absorption by ozone in the upper layers of the atmosphere), sufficient energy is available above this wavelength to degrade hundreds of grams per hectare per day of uniformly spread pesticide deposits, provided the quantum efficiency of the reaction is greater than 10 per cent[1]. Simultaneous exposure of pesticides to air and light can result in a range of chemical reaction types.

Hartley and Graham-Bryce[1] list a number of photodegradation reactions that have been reported. These include bond rearrangements, oxidations, ring fissions and elimination of, or substitution by, halogens. It should be noted that unsaturated lipids (e.g. terpenes) present in plant cuticles can shift the absorption maxima of organic materials by charge transfer and other electronic interactions, thereby protecting insecticides dissolved in the cuticular waxes from photodegradation.

Biotransformations by surface microflora also cause chemical degradation and hence loss of biological activity at the plant surface. Briggs[39] has ranked the various types of chemical bonds that characterize organic pesticides in terms of their chemical stability to attack by microorganisms. His data are based on biotransformation by soil microorganisms, but their application is probably more general and may well apply to microflora at the plant surface. Although such flora certainly exist on plant surfaces[40, 41], we could find no reports of biodegradation of insecticides by microorganisms taking place on foliage or stems. Examples of biodegradation in soil and within animals and plants are legion, however. Examples of organisms known to metabolize a fungicide and which occur on plant surfaces are to be found in the literature. *Chrysosporium pannorum*, which will degrade the organomercury compound verdasan[42, 43], was recently shown to colonize the foliage of *Agrostis tenuis* treated with the fungicide[44]. It therefore seems likely that insecticide deposits on plant surfaces provide a useful carbon source for microorganisms unaffected by their toxic action.

2.3 Transfer of insecticide to the insect surface

2.3.1 Transfer by surface contact

For an insecticide to exert its toxic action, enough material must be available on the plant to ensure that transfer to the insect will be sufficient to result in death. The extent of the transfer of material will depend both on the probability of an insect encountering a toxic residue and the proportion of insecticide which adheres preferentially to the insect cuticle. The probability of encounter will depend upon factors such as the fractional cover of the treated surface, the sites of deposition on the plant and the walking and feeding behaviour of the insect pest (see Sections 2.3.2 and 2.3.4). A theoretical study of the factors determining surface cover has been published by Johnstone, who points out that the increase in cover obtained by increasing droplet density decreases as the probability of superimposition of drops increases[45].

(a) Factors determining the amount of insecticide transferred to the target organism. The encounter probability of larvae of *Spodoptera littoralis* placed on oil-based ULV leaf deposits of permethrin has been described and modelled stochastically by Salt and Ford[20]. These authors considered that the quantity of insecticide encountered can be determined from the area swept by a larva as it moves over a treated surface, the density and diameter of the droplets, the proportion of the body making contact with the plant surface and the behavioural state (e.g. feeding, resting and walking) of the larva. Similar views have been expressed by Hartley and Graham-Bryce[1]. Salt and Ford[20] recognized that the kinetics of the pickup process and subsequent intoxication might be profoundly affected by the different modes of contact associated with each state. They concluded that in order to account for the time–dose–response relationships observed when larvae were placed on permethrin-treated cabbage leaves, the proportion of the available liquid deposit transferred by contact from the leaf surface to the larvae must be close to unity. The transfer process was therefore characterized by a high adhesion between the oily ULV deposit and the larval cuticle, a conclusion later confirmed by observations using light microscopy.

The extent of transfer of insecticide from plant to insect will be determined by the adhesion forces acting between the deposit, the plant (intermediate) and the insect (target) surfaces, and possibly by the internal cohesion of the deposit itself. These forces determine the effectiveness of retention by each surface once contact has been made. The competition for insecticide between the two surfaces results in a steady initial accumulation of toxicant by the organism, leading to the establishment of a steady state when the rate of pickup by the insect equals the rate of detachment[20]. Lewis and Hughes[46] studied the contamination of blowflies exposed to isometric particles of polar and non-polar dyes deposited on surfaces with differing properties. The effectiveness of retention by such surfaces was related to the interfacial tensions between the deposit and the competing surfaces, and their associated contact areas. The amount of dye transferred to the flies from dry filter papers increased linearly over an initial period of four minutes, whereas that from a paper impregnated with a hard plant wax rapidly approached a limit, consistent with the establishment of a steady state. Transfer was always more efficient for non-polar material — suggesting that lipophilicity increased adhesion to the blowfly cuticle — but was less marked for transfer from wax-treated papers. The oil-soluble particles presumably adhered more readily to the plant wax compared with the filter paper, so providing greater competition for dye. This interpretation is consistent with the observation that the rate of detachment of lipophilic particles was greater when flies were placed on wax-impregnated paper. Lewis and Hughes[46] also investigated the redistribution of accumulated dye over the insect surface. The immediate sites of contact and hence uptake, the tips of the ventral tarsal setae, were all available to act as reception points when the flies first alighted on a treated surface. Within seconds, however, insecticide was transferred to these sites and the number available for further reception fell, eventually to result in a net rate of pickup 0.25 times the initial rate. Detachment of particles commenced at an early stage, further reducing the net transfer to the insect but was diminished by the grooming activities of the treated flies. Grooming transferred a considerable amount of material from the setal tips to the more distal parts of the ventral tarsal setae and to the main tarsal cuticle, the basal parts of the setae, the head, the wings and the abdomen, where they were more likely to be retained.

The transfer of DDT particles from dry and waxy surfaces to the surface of adult blowflies was shown by Gratwick[47] to depend on similar factors to those of the dye particles of Lewis and Hughes[46], although when transferred from surfaces impregnated with plant wax DDT appeared to approach a steady state more slowly. Gratwick attributed a reduced rate of uptake of DDT, compared with dye particles, to the greater surface area of the needle-like DDT particles, which she suggested would adhere more strongly to the underlying surface[47]. Transfer of DDT from an oily surface was progressively impeded as the amount of oil on the surface or the time the particle was in contact with the oil increased. These results suggested that non-polar insecticides picked up by an insect with an oily cuticle, such as that of the cockroach (*Periplaneta americana*), would be retained rather than detached. Adhesion to oil appeared to be due to strong interfacial tension, since the viscosity of the oil and the dissolution of DDT in the oil appeared to have only a marginal affect on retention. The presence of wetting agent on a surface increased adhesion of DDT to that surface, presumably as a result of the increased area of contact and hence association with the underlying surface.

Adhesion of insecticide to the surface of insect cuticle has also been investigated by Armstrong *et al.*[48]. These authors measured the pickup of HCH by grain weevils (*Calandra granaria*) walking over filter papers impregnated with pure isomers ($11 \ \mu g/cm^2$). Their paper provides measurements that may be used to derive further information concerning the transfer of this insecticide (Table 2.6).

Table 2.6 Measured and derived values describing the capacity of the epicuticular wax layer of the grain weevil, *Calandra granaria*

Quantity	Measured value	Derived value	Units	Comments
Recovered epicuticular waxes	0.69		mg/(g body wt)	Chloroform extraction of 15 g of grain weevils yielded 30.8 mg of a brown amorphous material from which 10.3 mg of pure wax was harvested by fractional crystallization
Number of weevils per gram	500–600		Dimensionless	
Amount of epicuticular wax per weevil		1.15–1.38	μg/insect	
Average surface area of a weevil		7.85×10^{-2}	cm^2	Estimated by regarding a grain weevil as a cylinder 2 mm long by 0.5 mm radius, i.e. surface area $= 2\pi r^2 + 2l\pi r$
Average wax cover of the insect cuticle		14.6–17.6	μg/cm^2	
Approximate thickness of the epicuticular wax layer		0.1–0.2	μm	These values are similar to those obtained for other insect species

The results are in broad agreement with those of Lewis and Hughes[46] and Gratwick[47]. The insecticide that accumulated on the cuticle surface was in the approximate ratio of the solubilities of the HCH isomers in hydrophobic solvents including hydrocarbons, esters and ether. Penetration rates were similarly related to the solubility of the isomers in the epicuticular wax, suggesting that the rapid penetration of the γ and δ isomers was a consequence of a high local concentration of these insecticides at the insect surface. The accumulation curves describing the pickup of the γ and δ isomers are consistent with the establishment of a steady state between insecticide at the insect surface and that on the underlying filter paper. However, the steady state levels correspond to quantities of insecticide at the cuticle surface approximately equal to the saturation values of these materials in petroleum oil. This result suggests that steady state occurs when the epicuticular waxes are themselves close to saturation (Table 2.6). Thus, most of the accumulated toxin must be transferred by a process characterized by dissolution in the epicuticular wax. This process may involve vapour transfer from the treated filter paper[49].

The capacity of the insect epicuticular waxes will therefore have an important effect on the retention of insecticide by this surface, and also on the subsequent penetration rate. Approximately $20\,\mu g/cm^2$ of epicuticular wax (mass of wax per unit area of insect surface) was recovered by Armstrong et al.[48] following methanol washing of grain weevils (Table 2.6). Similar values have been reported for other insect species, including phytophagous pests: Hartley and Graham-Bryce[1] give a range of 0.05–$0.4\,\mu m$ for thickness of epicuticular wax, which can therefore vary considerably with species. This range is very similar to, although greater, than the range (0.1–$0.2\,\mu m$) of wax thickness found on leaf surfaces. Since both plant and insect cuticular waxes have a similar chemical composition[50–52], they should have similar affinities for insecticide. The capacity of each surface for insecticide can therefore be very similar, and suggests that following encounter between insect and deposit, the competition for the toxin will often be evenly balanced.

Transfer of toxin from residues dissolved in plant waxes will tend to be a slow process if it is based upon diffusion[53]. Most of the dissolved material is likely to remain in the plant wax, particularly if contact between the insect and the plant surface is fleeting. On the other hand, if sufficient insecticide is available above the plant waxes to allow transfer to the insect by adhesion rather than diffusion, pickup may be more significant. Deposits adhering to the cuticle but with profiles above the surface have been observed in scanning electron micrographs of insecticides deposited onto the surface of various insect species[54].

An increase in the surface cover by the insect waxes relative to that of the plant waxes will move accumulation of insecticide in favour of the insect: an increase in the plant wax layer relative to that of the insect will make retention by the plant waxes more likely. Such competition will be sensitive to small changes in the ratio

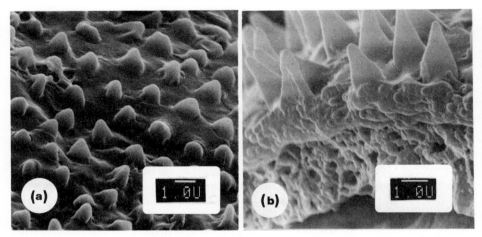

Figure 2.6. Scanning electron micrographs of the cuticle surface of second instar larvae (body mass 0.5 mg) of *Spodoptera littoralis* (Boisd.) showing papillate topography. (a) Cuticle surface of second segment just prior to second moult. (b) Cuticle surface near the base of a foreleg just after the first moult. Dimensions are indicated by solid bars ($u = \mu m$).

of the wax cover of the two surfaces. As the amount of wax on one surface increases substantially over that on the other, the layer of higher wax cover will tend to behave as if it were infinitely thick. Under these conditions, the mass of insecticide transferred to the thin surface layer may become negligible. Such considerations could be crucial in determining the optimum cover (mass per unit area) of insecticide for an effective residual deposit on a plant surface. A thick layer, for example, will require a greater deposit cover to achieve a given concentration in the plant or insect wax. Penetration of insecticide into the insect will establish a steady state in the insect epicuticle as material flows from the plant surface to the insect tissues beneath the integument. Variation in the amount of plant waxes will modify the concentration gradient between the two surfaces, and hence the rate of transfer from plant to insect. These theoretical considerations suggest that studies of the transfer of insecticide from the intermediate surface to that of the target organism are essential if the contact efficacy of contact insecticidal deposits is to be much improved.

Hartley and Graham-Bryce[1] have suggested that the area of contact between the deposit and the recipient surface will also be a dominant factor determining transfer of pesticides to target insects. Recent work with the scanning electron microscope (Figure 2.6) has revealed that the surface of larvae of *Spodoptera littoralis* is extensively papillate, resulting in an up to a twofold increase in surface area, depending on the stage of larval development. This observation may partly explain the very high adhesion between the larval cuticle and the oil droplets of permethrin, reported by Salt and Ford[20]. High adhesion must also depend on the similar physicochemical nature of the oil and of the cuticle surface, however. The hydrophobic nature of most insect cuticles has been recognized for many years. Brown[11], for example, contrasts the high contact angle (180°) observed for distilled water placed on the surface of insects with the lower values obtained for aqueous solutions (0.5 per cent) of polar organic solvents. He notes that whereas the extent of spreading is determined by the surface tension of the applied material, its rate is dependent upon viscosity. The latter property may therefore be important in determining the steady state levels of insecticide accumulated by individual insects walking over treated substrate. Salt and Ford[20] and Armstrong et al.[48] point out, however, that steady states result not only from reversible uptake (attachment and detachment) but also by the loss from the surface through penetration into the insect which can be quite rapid. Increased viscosity may modify the rate of penetration, and hence the steady state level, by reducing both the rate of change in the area of contact of insecticide with cuticle surface and the rate of entry by fluid flow through microscopic pores in the cuticle structure. Both processes are considered to determine the rate of entry of insecticides into treated insects[1].

Steady states are established during the transfer of a synthetic amorphous silica dust (Wessalon S) to adult grain weevils (*Sitophilus granarius* (L.)), from a variety of surfaces ranging in texture from smooth (vinyl or glass) to rough (hessian sacking)[55]. Higher steady state deposit levels were observed on insects exposed to smooth treated surfaces compared with those exposed to rough surfaces. Higher levels were also reported for dry dust accumulating on the weevil surface, compared with an aqueous suspension of dust, although as the steady state level increased differences were less marked. As the rate of attachment of dust increased, so did the rate of detachment. On flat surfaces, the treated insects were knocked over and spent time struggling on their backs in an attempt to right themselves. During these periods, large quantities of dust accumulated on legs, thorax and elytra, from where they were easily dislodged. The detached dust was then available for subsequent attachment, perhaps to the same individual. Gowers and Le Patourel noted that the LD_{50} values for dust application were inversely related to the relative amount of dust removed over a 24-hour period and that the rate of exchange of dust may be of greater importance in determining toxicity than the absolute amount of dust accumulated at steady state[55]. This interesting observation suggests that it is the act of contact with a dust particle that results in the abrasion of the cuticular wax and thus leads to the death of the insect through water loss. The number of sites of abrasion will then depend on the number of contacts an insect makes with dust particles.

(b) Influence of particle size and surface properties. The effectiveness of pickup is dependent upon the deposit size. Early studies by Hadaway and Barlow[56] established that the rate at which DDT was transferred to adult mosquitos (*Aedes aegypti* (L.)) walking over deposits (2–5 mg/m^2) decreased with time, until by 32 minutes a steady state was approached. The rate of attachment and hence the

steady state level reached a maximum value for a range of particle diameters 20–40 μm, with both smaller and larger particles being less efficiently captured. The most toxic was 10–20 μm, suggesting that small particles may penetrate the cuticle more effectively. Broadly similar results were obtained with tetse fly adults (*Glossina palpalis* (R-D)). Large particles were more easily detached from the cuticle than small, thus leading to a lower steady state level on the insect surface and hence a less toxic residue. Small particles (5 μm) tend to have a high adhesion to a surface, making their transfer to the insect more difficult. Moreover, on rough surfaces, particles less than 10 μm may be inaccessible to contact by mobile insects. Beesley and Chadwick[57] recently provided scanning electron micrographs demonstrating that although wettable powders and microencapsulated formulations lie above the surface of smooth, impervious materials such as glass, small particles on rough surfaces such as wood collect at the bottom of furrows where they are less available to the target insect.

Chadwick and Carter[58] and Chadwick[59] report that factors other than surface roughness affect the probability of an insect contacting a deposit. Dissolution or absorption of the active ingredient in the underlying substrate appears to be a major source of loss of availability of material on a surface. Vinyl tiles and paints, for example, provide a continuous surface into which non-polar organic insecticides may be absorbed. The low mortalities observed for susceptible houseflies exposed to residual deposits of DDT ($2000\,mg.m^{-2}$) on these surfaces are, in part, attributable to surface absorption[58]. This process is associated with a rapid loss of effectiveness for many insecticide formulations applied to surfaces. Evidence supporting this view has been reported for the pickup of non-volatile ULV oil deposits of permethrin from the surface of cabbage leaves (*Brassica oleracea*)[20]. Light microscopy revealed that a high proportion (> 0.9) of the available insecticide was rapidly transferred to larvae of *Spodoptera littoralis* on encounter, provided the droplets retained a distinct profile above the leaf surface. Oil droplets placed on 'non-waxy' leaf surfaces retained a stable profile for many hours, but on waxy surfaces they spread rapidly, mixing with the epicuticular waxes, and were then less available for transfer to the target insect[10, 20]. There have been reports of increased effectiveness of deposits of non-polar insecticides on plant surfaces kept in atmospheres of high humidity[60]. Investigation of deposits on glass surfaces[61] suggested that, at high humidity, insecticide was preferentially available and more readily transferred to walking insects.

The mode of locomotion of an insect can influence the extent and nature of insecticide transfer. Adults of most orders present only a very small area of cuticle (the tips of the legs) to the surface over which they walk and will in any case spend much of their time in flight. Some genera exude a fluid secretion of hydrophobic material to increase their adhesion to the surface over which they are walking by increasing their contact[62]. Crawling insects such as lepidopterous larvae present a much greater area and remain in continuous contact with the surface: they are therefore more likely to acquire a toxic dose.

The effectiveness of insecticidal deposits against crawling insects is greatly influenced by the deposit size and density and mass of active ingredient (Section 2.3.4). Johnstone[63, 64] has demonstrated that these variables may be controlled by the spray operator to achieve a more effective control under field conditions. He investigated the effectiveness of particle size over a range of diameters of 0–40 μm for a wettable powder formulation of carbaryl applied at ULV to cotton to protect against first instar larvae of *Pieris brassicae*. Deposits containing particles of 15–20 μm were most effective, a result attributed to the increased pickup and retention of these deposits by the feet and claspers of exposed larvae. As observed earlier for flies[56, 65] particles larger than 20 μm were readily detached whereas the smallest particles adhered firmly to the underlying surface, in this case the cotton leaf. Johnstone investigated the relationship between ULV application and response for deposit densities of up to $64 \, cm^{-2}$ for a wettable powder formulation of carbaryl. He concluded that below $4 \, cm^{-2}$, the mortality of *Pieris brassicae* was limited by the low probability of larvae encountering sufficient deposits to acquire a toxic dose, whereas between 4 and $64 \, cm^{-2}$ encounter was more likely and the expected cumulative log-normal response curve based on the distribution of tolerances was observed. The value of the observed response depended on the particular combination of particle size, deposit density and mass application rate employed. For a particle diameter of 5 μm and a constant mass application, a combination of high concentration (0.025 lb/acre) and low deposit density (0.25–1.0 cm^{-2}) was most effective, whereas for a particle diameter of 18 μm, a large number of deposits ($4 \, cm^{-2}$) of low concentration of the active ingredient gave higher mortalities. Johnstone suggested that the larger particles were less well retained by the leaf surface, but sufficiently well retained by the insect surface to result in a more effective transfer to the pest.

2.3.2 Transfer by direct impaction

Insects also come into contact with insecticides as a result of direct interception of airborne droplets, exposure to insecticidal vapour or ingestion of treated food material. The process of direct interception has been shown to increase the accumulation of insecticide by caterpillars of *Spodoptera littoralis* feeding on plant leaves, under controlled laboratory conditions[66, 67]. The response of larvae exposed to direct sprays of bioresmethrin or permethrin was always greater than that observed for larvae placed on the leaves after spraying the insecticide as a protectant. The increased response resulting from direct application of permethrin was always more marked at the higher droplet densities, consistent with the increased probability of scoring a direct hit on impaction. Further transfer of insecticide by contact with residual deposits must occur after spraying and the response surface for eradication may be interpreted as a sum of the response due to residual deposits and that expected for direct impaction alone.

The acquisition of airborne drops by insects has been studied by Wooten and Sawyer[68] who found that approximately 80 per cent of insecticide intercepted by flying locusts (*Schistocera gregaria*), was deposited on the wings by sedimentation and impaction: the head capsule was also a likely target for deposition. High air speeds increased accumulation, presumably as a result of increased impaction, and small drops were retained more efficiently than large drops. The beat of the wings resulted in a loss of some of the liquid by redispersion, with large drops (2000 µm diameter) approximately three times more likely to be redispersed than small drops (600 µm). Although the pickup of airborne drops by plant and insect surfaces (Section 2.2) is determined by similar processes, the geometry of the target will affect the surface distribution of the retained deposit. Reception of insecticide by walking insects as a result of direct contact by sedimentation and impaction will occur at very different sites from the points of contact (e.g. tarsal tips) when insects walk across treated surfaces[57]. This may result in different rates of penetration through the cuticle and hence different rates of intoxication. During the early stages of transfer before symptoms of intoxication can occur, insecticide acquired by interception is less likely to be lost by subsequent detachment, since it will be retained at sites not normally touching the underlying surface.

2.3.3 Transfer by gaseous contact

Most of the work describing the transfer of insecticide to target insects has concerned liquid or solid deposits. However, some insecticides applied to plant surfaces have vapour pressures sufficiently high to result in gaseous transfer. For this to be effective, conditions of still air should prevail around the deposit. At wind speeds normally observed within a crop canopy, laminar flow results in a layer of stagnant air several millimetres thick immediately above the plant surface[1]. A surface deposit of an insecticide with a vapour pressure of 10^{-2} Pa will saturate the adjacent air and establish a steady state gradient across the stagnant layer. The thickness of this layer will depend on the surface geometry of the leaf and its absorptive capacity for the insecticide vapour[69]. If the layer of vapour is sufficiently deep, small arthropods such as aphids, mites or whitefly will be subject to an atmosphere in which a pesticide can approach its saturation vapour concentration (SVC) in air. Provisional calculations[1] have suggested that a lethal dose of insecticide may be picked up by aphids within 1 hour of exposure to such an atmosphere. As the size of an organism increases, however, more of its body will be held above the vapour and the importance of this source of toxicity will diminish. The effectiveness of vapour transfer is therefore dependent upon the body size, its location and behaviour, the air speed at the plant surface, and the surface properties, vapour pressure and toxicity of the active ingredient. Physiological factors, such as the probability of the spiracles opening and closing, may also determine the effectiveness of vapour transfer.

Examples of this mode of transfer are occasionally found in the literature. Vapour effects have been observed[70] for phorate and disulfoton granules applied to field bean plants, and eggs of a noctuid moth were reported to respond after shorter exposure periods to insecticidal vapour than the larger larvae of later instars[1]. The vapour transfer of γ-HCH (vapour pressure c. 1.3×10^{-3} Pa at 20 °C) from impregnated filter papers to the surface of the grain weevil (*Calandra granaria*)[48] was further investigated by Armstrong *et al.*[49]. When direct contact was prevented by separating the insects from treated filter papers by inserting a clean paper between the target insect and the treated surface, γ-HCH was as effectively transferred as when direct contact was permitted. This result contrasts with that found for DDT (vapour pressure c. 25×10^{-6} Pa at 20 °C), where transfer was substantially reduced by separating the insects from the treated surface by this technique. Further evidence in support of vapour transfer from insecticide deposit to target insect has been presented by Abdalla[71] who reported responses for sessile whitefly larvae four days after treatment at different radial distances from the centre of approximately spherical ULV deposits of permethrin or pirimiphosmethyl dissolved in a non-volatile oil. The response of larvae kept in still air decreased with increasing radial distance to describe a sigmoid curve that reached a minimum response at a measured distance from the deposit centre. For the non-volatile insecticide permethrin (vapour pressure c. 4.6×10^{-5} Pa at 30 °C) response was restricted to a distance of 0.7 mm. Response was evident at up to ten times this distance (7 mm), however, following application of the volatile material pirimiphosmethyl (vapour pressure c. 1.4×10^{-2} Pa at 30 °C), even though it is substantially less toxic to insects than the pyrethroid[72]. Differences in the mass flow of these materials from the deposit and over the leaf surface were too large to have been attributed to the spread of carrier. Measurements for non-volatile oil formulations on the surface of leaves of tobacco (*Nicotiana tabacum*) suggest that between 0.5 and 48 hours after spraying, droplets move only 0.007 mm[71]. The observed response to permethrin could be detected at distances (0.7 mm) consistent with diffusion of the active ingredient through the epicuticular waxes, assuming they behave as viscous oils or solids (Section 2.3.5). Abdalla has suggested that the large distance (7 mm) over which pirimiphosmethyl exerted its effect arose from gaseous diffusion in the unstirred layer of air immediately above the leaf surface. In order to test this hypothesis, he repeated the experiment, having physically isolated the deposit from the target by removing a section of leaf between the site of the deposit and the test insects. A curve relating mortality to radial distance was obtained that was identical to that initially observed for the continuous leaf surface, confirming the view that movement of the active ingredient through the waxy layer of the leaf surface was not responsible for the observed zone of control. When air was blown continuously over the leaf surface, however, the radial distance for mortality was reduced and more closely resembled that for permethrin, demonstrating that gaseous transfer of the insecticide was taking place under conditions of still air. These results clearly demonstrate the impact that vapour phase transfer may have on the area of leaf around a deposit.

2.3.4 Transfer by feeding

For phytophagous insects, insecticide can enter the body via the cuticle (contact action) or the gut (stomach action). These two routes can be difficult to distinguish. Since it may take some time for a compound to penetrate to the site of action on contact, an insect may feed and ingest poison for some time before 'contact' symptoms begin to appear. Thus, both routes will be utilized and penetration will occur simultaneously through the insect cuticle and the gut wall.

Although ingestion of contaminated food by phytophagous pests has long been recognized as an important route for insecticidal action, detailed investigations of the mechanisms of transfer via the gut are difficult to find. Stomach poisoning is probably most important when the insecticide deposit has dissolved in the plant wax, and is then less available for transfer to the target insect by surface contact. Evans[73] has presented evidence to show that pickup of a dry deposit of permethrin, presumably lying upon the leaf surface, was approximately 50 times more effective when acting as a contact insecticide than when presented as a stomach poison. Price observed that cypermethrin was effective as a stomach poison when applied in a volatile carrier (aqueous acetone (10 per cent v/v) containing the surfactant Triton $\times$ 100 (0.025 per cent)), although more recent studies have indicated that it, too, acts primarily as a contact insecticide[74]. The presence of a volatile solvent such as acetone as a carrier has been shown by Stevens[75] to result in very rapid penetration of a pesticidal solute into the epicuticular waxes of a variety of plants. Since insecticides applied in volatile carriers or as emulsifiable concentrates often mix intimately with epicuticular waxes, they are more likely to be transferred by ingestion than wettable powders or non-volatile oils, which remain proud of the leaf surface following deposition, for which contact action is favoured.

If different routes of entry can result from different modes of application, then deposit properties may be adjusted to achieve selectivity. It has been suggested, for example, that the movement of material from the surface and its associated accumulation and retention within the leaf, where it is available for ingestion, is responsible for avermectin B_1 having less impact on populations of beneficial insects compared with plant feeders[76]. Adjusting application parameters can modify the probability of an insect encountering the discontinuous insecticidal deposits that result from ULV application. Salt and Ford[20] compared the velocity of movement of larvae of *Spodoptera littoralis* across a leaf surface for the two behavioural states, walking and feeding, and concluded that the proportion of the contacted deposit transferred to the insect may vary with behavioural mode. Unless repellency occurs, transfer efficiency is always likely to be high for ingestion, but more variable when contact is the result of walking. The average walking velocity (0.45 cm/s) of larvae of *S. littoralis* is 45 times faster than that for feeding (0.01 cm/s). Assuming approximately equal transfer efficiencies on contacting deposits by either behavioural mode — a reasonable assumption for involatile oils deposited on relatively polar leaf surfaces[20] — the probability of contacting and transfering insecticide at low deposit density will be determined by the ratio of the walking and feeding velocities (45:1). The value of this ratio is similar to the values of the ratios of LD_{50} values obtained by Evans[73] for permethrin and by Price[74] for cypermethrin applied as a contact and stomach poison against *S. Littoralis* larvae. This observation suggests that pickup may be directly proportional to the area of leaf contacted by larvae per unit time, which in turn will be determined by the feeding or walking velocities.

2.3.5 Influence of insect behaviour on the transfer process

Although the properties of the insecticide deposit can have important consequences for the transfer of the active ingredient to the target insect, the behaviour of the insect is also crucial, as we have indicated in the preceding section. The factors determining the behavioural responses of insects to insecticide, however, are not well understood and general principles have not been fully established. This is reflected in the small volume of published work that examines the influence of insecticides on the behaviour of the target insect, suggesting that this is an area of research deserving of further study. Behaviour may be modified by the very presence of insecticide in a manner that may influence the transfer process. Feeding behaviour, for example, can be characterized by a number of identifiable steps that include initiation, continuation and cessation. Initiation is probably prevented by *repellents*, continuation by *antifeedents*. Ruscoe[77] has reported antifeeding behaviour of caterpillars exposed to sublethal doses of permethrin. It is not clear whether this effect is strictly antifeeding or repellancy: the insect may have contacted sufficient material to induce sublethal symptoms of intoxication.

There have been a number of investigations confirming the antifeeding properties of sublethal doses of pyrethroid insecticides[78, 79]. In a detailed study[80] of the antifeeding effect in *Drosophila melanogaster*, a reduction in food consumption by flies treated topically with 1.25 ng of permethrin per insect (LD_{10} = 7.85 ng/insect) was related to a reduced frequency of the commencement of feeding. Once feeding was under way, however, its duration was unaffected by this low dose of toxicant. Permethrin applied at this dose also reduced the frequency of tasting which usually preceded feedings, suggesting that the antifeeding effect results from a loss of ability to sense food, rather than repellency. A reduced sense of taste may be an indirect result of an irritation response, rather than a direct effect of the insecticide on the mechanism of feeding. Sublethal doses of pyrethroid may affect a number of other biological activities, presumably as a result of an increased irritability induced by the insecticide[78, 80]. These include an increased frequency of commencing preening, which can result indirectly in a reduced length of the period spent in locomotion[80]. Our own observations of larvae of *S. littoralis*[20] suggest that individuals only react to droplets of permethrin in a non-volatile refined hydrocarbon when they contact the deposit. Symptoms of intoxication then usually follow in a short time. Symptoms of irritability such as hyperactivity, incoordination and involuntary body movement will increase the probability of encountering further deposits, particularly when large larvae are exposed to high droplet densities. However, Antunes de Almeida[81] reports that larvae of *Plodia interpunctella* avoided contact with drops of Shell Risella oil containing natural pyrethrins. The effect increased with increasing insecticide concentration. Some avoidance behaviour was also observed for drops of carrier oil containing no active ingredient. Avoidance was most marked for oils of low boiling point: non-volatile oils, however, behaved similarly to controls.

Being non-volatile, permethrin is less likely to be detected by the olfactory organs of insects. Volatile materials, however, should exert their effects (avoidance or repellency) over greater distances. This statement is based on the assumption that the very low vapour concentrations associated with low vapour pressures (Table 2.5) will result in an exposure to the insecticide too low for repellency. The repellent nature of the pyrethroids, deltamethrin and fastac, to honeybees[82, 83], however, suggests that if the insecticide is sufficiently active on sensory receptors, avoidance of deposits may occur even at very low vapour concentrations. Even non-volatile insecticides such as DDT can induce an alarm response, perhaps because they stimulate the production of pheromones. Demeton-*S*-methyl and pirimicarb were recently reported to induce alarm pheromone release in the aphid *Myzus persicae*, causing rapid dispersal of individuals[84].

Avoidance behaviour was also exhibited by tobacco budworm larvae placed on malathion-treated tobacco leaves[85]. Third instar larvae appeared to sense the presence of insecticide deposits. The larvae often stopped and raised and moved their heads sideways prior to avoiding a deposit. Large mature larvae or third instar larvae that had previously made contact with a deposit showed no tendency to avoid drops and moved rapidly in one direction colliding haphazardly with successive deposits. Avoidance response on an insecticide-treated surface is probably more important for small insects (1–20 mg) which can follow a route between small particles (100 μm diameter) without touching insecticide. Deposits of this size may be perceived as relatively large obstacles and easily avoided. Large larvae (100 mg body mass) will have greater difficulty in avoiding deposits of the same diameter (100 μm), however, since their ambulatory appendages will be much larger than these deposit dimensions. The situation will then be akin to the reader walking across a gravel path avoiding high densities of chippings of a given size or colour — a rather difficult exercise.

Deposits of insecticides may cause irritancy to exposed insects which then escape before acquiring a lethal dose. Exposed individuals may be more likely to take avoiding action on encountering further deposits of the same material. Miall and Turner[86] have suggested that this sequence of events will reduce the efficacy of a deposit in controlling a pest. They describe laboratory tests to determine the importance of irritancy of residual pyrethroids to German cockroaches (*Blattella germanica*), and the probability of avoiding action by the insect when the opportunity for escape was available. Irritancy and avoidance only occurred at low levels of accumulation but as the likelihood of transfer increased, irritant effects became unimportant. Repellency also resulted in increased survival of the rust-red flour beetle (*Tribolium castaneum*) exposed to insecticide-treated surfaces[87]. When individuals were placed in half-treated arenas, wettable powders were always more effective than emulsifiable concentrates, an effect shown by scanning electron microscopy to be due largely to the high adhesion of solid particles to the insect cuticle. A wettable powder blank containing no active ingredient, however, was significantly repellent over the first few days of exposure, wheras an emulsifiable concentrate blank was innocuous. Insects could detect differences between the two halves of an arena treated at different dosages (particle densities) of a wettable powder blank, with most individuals resting in the lower-dose-treated arena. Wildey[87] has suggested that exposed individuals can learn to distinguish between these treatments. Evidence for learning ability in insects has recently been presented by Hoyle[88]. Such responses are not observed for all insects, however. Gowers and Le Patourel[55] failed to observe any response by the grain weevil (*Sitophilus granarius*) similarly exposed to silicaceous powders used to control insects by abrasion, although Ebeling *et al.*[89, 90] and Rust and Reierson[91] have described examples of avoidance of a variety of insecticidal deposits by species of cockroaches.

2.4 Relationship between spray application characteristics and biological response

Application characteristics can be adjusted to modify the response of insect populations. Population response to treatments applied under field conditions is not, however, easy to predict. A large number of variables operate in the field, many of which are difficult or impossible to control. When embedded within such heterogeneity, the specific effects of insecticide treatment on the dose received by individual insects, and hence the response of the population, can be difficult to identify. While multivariate methods of analysis may assist in identifying the more important components of this variation, the results of such methods are often difficult to interpret.

2.4.1 Laboratory studies on mobile pests

A number of workers have attempted to avoid these problems by simulating field treatments under controlled laboratory conditions. This has sometimes been followed by trials to determine whether similar trends can be identified in the field. Laboratory studies by Ford *et al.*[66] have demonstrated (Figure 2.7) large

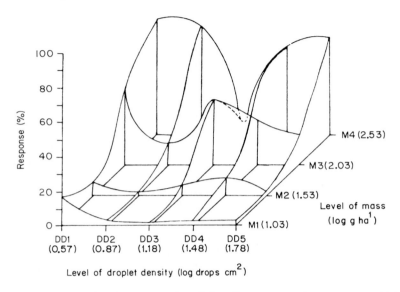

Figure 2.7. Three-dimensional representation of the effect on percentage response of late instar *Spodoptera littoralis* larvae resulting from the interaction of mass (log g/ha) × droplet density (log drops/cm²) and bioresmethrin applied as a residual deposit.

differences in the response of *Spodoptera littoralis* larvae to bioresmethrin on leaf surfaces for the same mass application rate applied at different volumes with a fixed distribution of in-flight droplet sizes (approximately 80 ± 10 μm diameter). This result has important implications, not only for optimizing efficacy but also for understanding the manner in which insecticides behave on plant surfaces. Using a factorial design, they investigated the effect of different combinations of mass and volume application rates on the proportion of larvae knocked down and killed at various times post-treatment. The concentration of active ingredient was varied to produce droplets containing different masses of insecticide, which were applied at different deposit densities. When investigated as a residual deposit, a spray pattern comprising a few deposits each containing a high mass of active ingredient was generally more effective than one made up of many deposits each with a low mass of insecticide (Figure 2.7). This suggests that, at low concentration, the probability of transfer of pyrethroid to the insect is more critical than the probability of encounter of a deposit by a larva. However, the situation is more complex than this. At a high mass application rate, an increase in the droplet density (a decrease in the mass per droplet) resulted in an initial decrease in the response of the insect, which fell to a minimum at approximately 10 drops/cm². Thereafter, the response increased, suggesting that an increased probability of encounter was overcoming a decreased amount of insecticide available to the insect. At a low mass application rate, such an increase in response did not occur with increased deposit density. These observations imply that the probability of transfer may be related in a non-linear manner to the mass of active ingredient per deposit.

Ford *et al.* suggest that there is a critical mass of insecticide per deposit: below this value, insufficient transfer can occur from leaf to insect to result in a response[66]. Although the critical mass itself may vary with droplet density, and hence be influenced by the probability of encounter, they suggest a value for bioresmethrin on spinach beet (*Beta vulgaris*) of 1–3 ng per deposit. The mean in-flight diameter of the ULV drops was estimated as approximately 80 μm; thus, assuming a spread factor on impaction of approximately 2[1], the area covered by the average bioresmethrin deposit is 1×10^{-4} cm^2. Insecticide deposit of less than 10–30 μg/cm^2 is therefore unlikely to result in sufficient transfer of insecticide to kill 100 mg larvae. Similar values have been reported for the wax cover of many plant surfaces[1, 19], suggesting that transfer from leaf to larva may be prevented by dissolution of insecticide in the wax. If the deposit cover falls below the plant wax cover, complete dissolution is possible and insecticide will penetrate into the underlying wax layer where it is unavailable for efficient transfer. If the deposit cover is substantially in excess of that of the plant wax, the underlying wax will become saturated and only a small proportion of material will be lost from the surface by diffusion into the leaf surface[60]. As has been shown previously (Section 2.2.3), a non-polar material such as bioresmethrin is unlikely to penetrate very far into the underlying cell walls which comprise polar materials. Because the rate of diffusion of dissolved material in the solid cuticular wax should become negligible within a few tenths of a millimetre of the deposit edge, the plant cuticle will have a finite capacity for the insecticide related to the thickness of the wax layer. As the mass of insecticide per unit area is increased above that of the plant waxes, the insecticide deposit will be retained proud of the leaf surface where it will be more readily available for transfer to a walking larva.

Reay and Ford[67] obtained similar results for the persistent pyrethroid, permethrin. The differences in response associated with various combinations of droplet density and mass of insecticide per deposit, however, were less than those observed for bioresmethrin. Since permethrin has a similar toxicity to bioresmethrin when applied topically to larvae of *Spodoptera littoralis*, Reay and Ford suggest that this result might have arisen from the greater persistence of permethrin on the leaf surface, resulting in an increased probability of encounter by larvae due to the extended life of the deposit. Subsequent laboratory and field trials[92] have confirmed that response is dependent upon the mass of insecticide per deposit and the number of deposits per unit area of leaf. A combination of high masses per deposit and low droplet densities were most effective on cotton, whereas a larger number of deposits each containing less active ingredient gave maximum response when applied to lucerne, suggesting that the behaviour of contact deposits varies with plant surface. For maximum efficacy against mobile insects, the application characteristics of contact insecticides must be suited to the attributes of a particular crop. With systemic

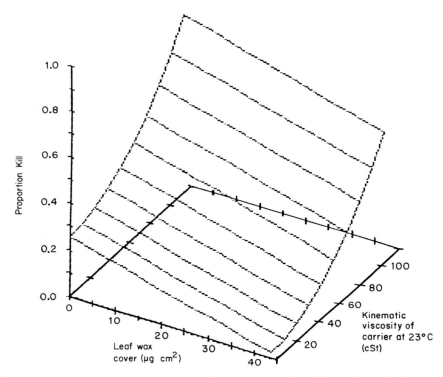

Figure 2.8. Response surface for early fourth instar larvae of *Spodoptera littoralis* (mean body weight = 14 mg) placed on cabbage leaf surfaces treated with cypermethrin (5g/l) at ULV using Sirius oil carriers of different viscosity (80 μm drops; 5 drops/cm²).

insecticides, the situation will be different, however, since encounter and transfer probabilities are determined by different processes.

Formulations that keep the active ingredient proud of the plant surface should result in increased efficacy. Viscous carrier oils that spread slowly and maintain high contact angles and convex profiles (Section 2.2) should therefore be transferred efficiently to the target insect. Efficient transfer will be opposed, however, by the waxy nature of many leaves, particularly if the wax deposits are characterized by bloom: this has been confirmed by studies in this laboratory. Larvae of *Spodoptera littoralis* were placed on cabbage leaf surfaces treated with cypermethrin to give solutions of fixed concentration applied at constant droplet density and drop size. Larvae were placed on this surface 24 hours after spray treatment and mortality observed 48 hours after placement. The response surface for different combinations of carrier viscosity and leaf wax cover (Figure 2.8) determined by regression analysis shows that, whereas the proportion killed falls linearly with increasing wax cover, there is a non-linear increase with carrier oil viscosity[10]. This result suggests that a considerable increase in the efficacy of oil-based ULV formulations for the control of moving pests in the field should be expected, provided deposits are retained above the plant surface and are able to withstand weathering.

2.4.2 Laboratory studies on sessile pests

An interesting study of the influence of spray properties on the response of the eggs of the glasshouse red spider mite (*Tetranychus urticae*) has been reported by Scopes[93], Munthali[94] and Munthali and Scopes[95]. Uniform drops of the acaricide dicofol were applied to cover the adaxial surface of bean leaves (*Phaseolus vulgaris*) infested with known densities of eggs. The concentration of active ingredient and the size of the droplets were varied, and mortality recorded 4 days post-treatment.

Very few eggs were hit directly by spray droplets: for example, an application of 200 drops/cm^2, each with an in-flight diameter of 53 µm, resulted in less than 10 per cent of the eggs being hit on impaction. Biological activity must therefore rely on subsequent spread of acaricide, although the authors could not determine directly whether such movement took place over or through the wax of the leaf surface. Response resulting from a particular treatment was observed to vary between leaf types. Estimates of the LD_{50} (µg active ingredient/cm^2 leaf surface) of dicofol were obtained for a variety of treatment combinations of concentration and drop size. Because of the even distribution of eggs over each leaf, Munthali and Scopes[95] suggest that the LD_{50} represents the mass of acaricide necessary to render 50 per cent of the leaf area toxic to the target. If this assumption is valid, the bioassay may be used to compare how the differential spread of pesticide is affected by concentration, drop size, formulation and leaf type.

Munthali[94] developed the concept of 'biocidal area' which he defined as the area of control per deposit at $p > 0.5$. Although this is a useful method of expressing the effectiveness of a treatment in terms of the leaf cover, it is not based directly on the unit mass of active ingredient per deposit. A related quantity which we term the 'biocidal efficacy' of a deposit may be obtained by inverting the LD_{50}. This term will have units of area per mass of active ingredient and describes the effectiveness of a treatment at $p > 0.5$: it can be considered to represent the area of 50 per cent control achieved per unit mass of pesticide.

Munthali's results are presented in Figure 2.9. An optimum concentration for maximum effectiveness at a fixed drop size is indicated, although a marginal decrease in the value of this optimum may occur for an increase in droplet size. The efficacy observed at the optimum concentration, however, shows a marked increase with decreasing drop size. These results have considerable implications for the choice of the most effective spray characteristics. Paradoxically, the most effective treatment is not achieved at very high concentration but at an optimum value of concentration dependent upon drop size, and consequently droplet density. The cover of the leaf surface by acaricidal deposits must presumably be

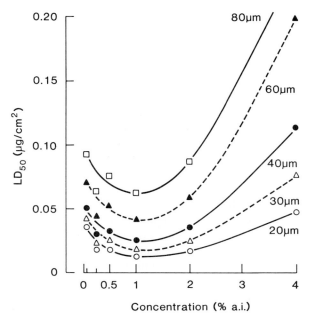

Figure 2.9. The biological effectiveness of dicofol applied at ULV to the adaxial surface of bean leaves (*Phaseolus vulgaris*) infested with eggs of the glasshouse red spider mite (*Tetranychus urticae*). Killing potency estimated as LD_{50} four days post-treatment is a function of the concentration of active ingredient in the formulation such that an optimum concentration for maximum treatment effectiveness can be identified for a given droplet size (μm).

varying between treatments in a manner that modifies the effectiveness of transfer of pesticide to the target eggs.

The paradox can be understood when it is realised that biocidal efficacy depends not only on the distribution and transfer of dicofol from acaricidal deposit to egg, but also on the tolerance distribution of the eggs. Movement of acaricide from a spread deposit to an egg located at a fixed radial distance from the deposit centre is probably a diffusion-dependent process. The concentration of active ingredient will therefore decrease with radial distance from the deposit centre to produce a bell-shaped concentration profile with an extended tail. As time elapses, the insecticide will be transferred from the initial deposit to cover an increasing area of leaf surface, but the rate of diffusion will progressively decrease. The rate of movement will then be so slow as to effectively preclude further distribution of pesticide and a stable zone of influence will be established. The exposure of an egg to acaricide will therefore be a function of its distance from neighbouring deposits, the diffusion coefficient of the insecticide on or in the epicuticular wax and the elapsed time after impaction. Although increasing the concentration of active ingredient may have a significant effect on rate of diffusion prior to the establishment of non-equilibrium, its effect thereafter becomes negligible and is unlikely to result in a significantly larger zone of control.

Table 2.7 Biocidal efficacies of various stages in the life cycle of the whitefly (*Trialeurodes vaporariorum*)[94]

Stage	Biocidal efficacy ($p \geqslant 0.5$) ($cm^2/\mu g$)
Egg	13.81
First instar	3.93
Second instar	1.41
Third instar	0.95

Increasing the spray concentration simply builds up the mass of active ingredient at the centre of the biocidal area of each deposit.

The cumulative log-normal distribution of tolerances of individual eggs to different concentrations of dicofol, in common with other pesticides, will tend to follow a sigmoid curve, with an upper limit corresponding to complete control of the population. This limit will fix a maximum concentration necessary for complete control and exceeding this dose cannot result in further response. Increasing the spray concentration will initially produce a more effective treatment, as the biocidal area of each deposit increases prior to the establishment of a stable zone. Once this zone is achieved, however, the biocidal area cannot be increased further, and further increases in mass of active ingredient per deposit will be wasteful. Thereafter, efficacy will only improve as droplet density is increased (by increasing the volume of application), so achieving a more continuous distribution of pesticide on the leaf surface. A theoretical basis for the maxim, 'if it doesn't work — use less'[96], is therefore established. In practice, it may be necessary to increase the concentration further to ensure adequate persistence.

Palmer *et al.*[97] used the procedure adopted by Munthali[94] to compare the effectiveness of sprays of permethrin of different drop size and concentration to control immature stages of whitefly (*Trialeurodes vaporariorum*) present on leaves of tobacco (*Nicotiana tobacum*). The authors expressed their results as LD_{50} ($\mu g/cm^2$), which may be converted to biocidal efficacies (see Table 2.7). Early, immature stages with small body mass are controlled at greater distances from the deposit centre and hence over greater areas than larger forms. Transfer of sufficient insecticide from the leaf to the insect to kill small individuals must therefore be possible at distances where the levels of active ingredient on the leaf are too low to cause this response in larger organisms. Reducing the mass of active ingredient per deposit results in a greater increase in the biocidal efficacy of treatments comprising many small drops compared with those based on fewer, larger drops (Table 2.7). This observation suggests that, provided the mass of active ingredient per deposit is sufficient to maintain a desired level of control, the effective cover of the leaf by insecticide is increased by a deposition pattern with a

Table 2.8 Biocidal efficacies ($cm^2/\mu g$) for permethrin deposits applied to tobacco leaf surfaces as formulations based on solvents of different physicochemical properties[90]

Formulation	Solvent composition	In-flight droplet diameter	
		50 µm	80 µm
JF8131	Volatile polar solvent plus volatile paraffin	0.31	0.25
JF8130	Low volatile chlorinated plus paraffinic solvents	0.95	0.27
JF8132	Volatile polar solvent plus low volatile paraffin	1.03	0.21
JF8133	Volatile polar solvent plus low volatile paraffin plus very low volatile paraffin	1.19	0.21
VK1	Very low volatile polar solvent plus low volatile paraffin	2.43	0.75

large cumulative deposit perimeter and hence diffusion front. Tenative results consistent with this view have recently been described[98, 99].

Palmer *et al.*[97] have also shown that the biocidal area of a drop can vary from 0.195 to 0.508 mm^2 depending on formulation. This result probably reflects differences in the initial spread of the impacted droplets, which will vary with the viscosity and surface tension of the formulation[1, 10, 11]. Increased efficacy should therefore be obtained by matching the surface activity of the deposit with that of the leaf[19]. Biocidal efficacies (Table 2.8) have been calculated from the data of Scopes[93] for permethrin deposits produced by formulations containing different carrier solvents and applied at a concentration of active ingredient (10 per cent w/v). The composition of the formulation clearly determines the area of control that can be achieved per unit mass of insecticide. The biocidal area may be considered to comprise two zones: an inner zone corresponding to the initial area covered by the spread droplet and an outer annulus that comprises insecticide that has diffused through the epicuticular waxes from the margin of the spread drop. Changing the formulation by adjusting solvent composition will alter the viscosity and surface tension of liquid deposits and hence the area of the leaf covered by the inner zone. Data in Table 2.8 suggest that volatile solvents that evaporate prior to impaction give rise to solid residues for which surface spread may be negligible. Such formulations will probably result in small inner zone areas and hence small biocidal efficacies. Non-volatile solvents will retain their liquid character and, providing their chemical composition is similar to that of the underlying leaf surface, will spread to increase both the inner zone and biocidal efficacy. As has been noted (Section 2.3.1), the more closely a liquid deposit resembles the underlying leaf surface the greater the extent of spread. Results suggest that tobacco leaf surface is best

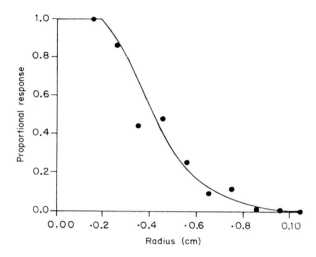

Figure 2.10. Mortality of whitefly larvae as a function of the radial distance from the centre of a ULV deposit containing 10 per cent w/v experimental formulation of permethrin (ICI Plant Protection PLC) applied to the surface of infested tobacco leaves. Time after application = 4 days, in-flight drop diameter = 114 μm, diameter spread factor = 1.7.

matched by a mixture of polar and non-polar solvents (Table 2.8): low volatility is important if the liquid state of the deposit is to be maintained.

The insecticidal efficacy of a deposit is likely to be influenced by the length of the deposit boundary per unit mass of insecticide retained within the deposit. This quantity is a measure of the effective diffusion front from which active ingredient can flow. Evidence to support this view is presented in Table 2.8. These results show that biocidal efficacy was always greater for the smaller 50 μm drops. This result suggests that diffusion of active ingredient will be most effective from a population of drops of small radial diameter with a proportionately large total drop perimeter. Small drops will produce annular diffusion zones that form a greater proportion of the biocidal area than will be obtained with large drops. The toxic residue will then cover a larger area of plant surface, so increasing the probability of encounter.

In another set of experiments[94], counts of whitefly larvae mortality were made at varying distances from the centre of permethrin deposits formed from impacted droplets of known in-flight diameter (Figure 2.10). At high spray concentrations, the proportion of larvae dead at the centre of the deposit was unity. The upper response limit ($p = 1.0$) at the centre was maintained for a radial distance probably corresponding to the diameter of the spread deposit after impaction. At greater distances, the proportion responding decreased to approach a lower limit characterized by an extended tail: this limit corresponded to the natural or control response. Four days after treatment with permethrin, low levels of response could be detected at a distance approximately 1 mm from the

Table 2.9 The effect of spray concentration on the movement of insecticide from the initial deposit

	Concentration		
	5%	10%	20%
In-flight drop diameter (μm)	117	114	113
Effective diameter for 50% mortality (μm)	650	830	920
Relative radial distribution[a]	5.56	7.28	8.14
Biocidal area relative to that for 5% spray concentration[b]	1.00	1.72	2.15

[a]Measured as the ratio of effective diameter to in-flight diameter.
[b]Biocidal area measures the area for 50 per cent control around an insecticide deposit.

deposit centre (Figure 2.10). With droplets of similar size in flight, the active ingredient moved further from the deposit as the spray concentration increased (Table 2.9). However, a fourfold increase in the concentration of the active ingredient produced only a 2.2-fold increase in the biocidal efficacy, suggesting that biocidal area is approximately proportional to concentration$^{2/3}$. This observation suggests that although the settled drop gives a three-dimensional deposit whose mass is proportional to the concentration of active ingredient, the effective distribution of insecticide and hence its biological efficacy tends to increase only in proportion to the leaf area occupied. Thus, a fourfold increase in concentration (from 5 to 20 per cent) increases the biocidal area by little more than twofold. This example of the law of diminishing returns is consistent with the spread of insecticide being diffusion controlled.

2.4.3 Field studies

As we have already noted, the effect of insecticide treatment of field crops can be difficult to identify and assess because of the level of observed but unattributable variation. Such variation may, however, modify the effectiveness of a treatment.

Uk and Courshee[100] report a field trial in which the observed distribution through the crop canopy of profenofos applied at ULV was related to the response of a hypothetical insect pest population. They derived an analytical model to relate the distribution of deposits in a crop to biological effectiveness. Consider a given application[1] and set of spray characteristics. Each deposit density class (D) on leaves will have a probability $P(D)$ of occurrence in the crop. If the probability of response of a target insect to a deposit class D of an insecticide is $Q(D)$, and $P(D)$ and $Q(D)$ are associated with independent events, then the probability $E(D)$ of a biological effect of that spray deposit is given by $E(D) = P(D)Q(D)$. The authors argue that the variability of the deposit density in the field would therefore tend to result in an overall lower biological effectiveness than if the chemical had been evenly deposited. For the case of a sedentary pest which is equally likely to occupy any part of the spray canopy and a contact pesticide with no other route of entry, an effective field dose f_c to produce a given percentage response can be estimated from the cumulative distributions of $P(D)$ and $Q(D)$, with deposit density as

$$f_c = \sum_{i=1}^{n} E(D_i)$$ (2.4)

where there are n classes of deposit density. $Q(D)$ may be estimated from laboratory data and analysed by probit or logit analysis and $P(D)$ from field trial data describing the distribution of insecticide through the crop canopy.

The authors cite a hypothetical example in which the maximum f_c is approximately 0.38 in contrast to an expected laboratory mortality $Q(D)$ of 0.96. This deviation of the field effect f_c from the laboratory response $Q(D)$ is caused by the inherent unevenness of the deposits in the crop canopy described by the distribution of $P(D)$. It is important to note that the bias invariably acts to diminish the field effect with respect to the laboratory response. Moreover, the bias will increase to result in an even greater diminution of effect if other sources of variation, such as avoidance behaviour by insects or preferential deposition on upper leaf surfaces, are present. Any application or formulation procedure that reduces the variation of deposit density throughout a crop canopy must therefore increase the field effectiveness. Such considerations justify the combined use of controlled droplet application (CDA) and electrostatic techniques of pesticide application[3, 4, 96].

Uk and Courshee[100] also consider the effect of increasing the mass of application on the expected field effect for their hypothetical data. They point out that for static chemicals acting as stomach and contact poisons, the upper limit representing complete mortality can only be approached gradually even though $P(D)$ values increase sharply. The result is that high levels of mortality in the field will only be achieved at relatively high dosage rates. This conclusion is based on the assumption that the slope of the deposit density distribution function for $P(D)$ values remain constant, whereas the location of the function (e.g. the mean, the mode or the median) will increase with the amount of insecticide applied. In their example, doubling the application rate increased the maximum f_c from 0.38 to 0.62, but in order to achieve an acceptable maximum f_c value of 0.95 a tenfold increase in the application rate would have been necessary. Laboratory experiments, however, would have suggested that only a sixfold increase was required.

An interesting set of data comparing the field performance of various pesticide deposits produced by electrostatically charged rotary atomizers with those released by conventional hydraulic sprayers has recently been obtained [101–106]. Although this work concerns the application of herbicides and fungicides as well as insecticides, the results summarized below show how the method of application can affect the biological response of a target organism under field conditions.

1. Greater amounts of active ingredient were deposited on plant surfaces following the use of electrostatic sprays, often resulting in high levels of control (above 90%) at reduced application. However, increased electrostatic deposition sometimes gave levels of biological response similar to those obtained with the lower deposits which result from the use of hydraulic sprayers. This may reflect the relatively shallow slope of the response curve as it approaches the upper limit corresponding to complete mortality.

2. Although higher deposit levels were generally obtained on exposed foliage and stems, inadequate penetration into the crop canopy sometimes leads to reduced effectiveness.

3. Uniform spacing of electrostatically applied deposits should result in better plant cover and, according to the argument of Uk and Courshee[100], better control. Because of the restricted movement of active ingredient from a deposit, cover will probably increase as drop size decreases and the number of deposits increases. The retention of small drops by plant surfaces, and hence the extent of insecticide cover, is likely to be enhanced by the use of charged spray particles.

4. Biological response depends partly on the position of the deposits on the plant surface. The use of electrostatic sprays permits both the charge per surface area and drop size to be adjusted to obtain a desired trajectory, and hence position on the crop. If a variety of deposition sites is required, charged sprays of variable-sized droplets, and hence charge to mass ratios, may be preferred to sprays of uniform droplet size.

5. Increased penetration into the crop canopy results from either vertical displacement of charged drops from the spray head or wind assistance of airborne droplets.

Several field studies have demonstrated that the use of concentrated sprays comprising small drops can be extremely effective[92, 95, 107], particularly when used at low mass application. Other cases are reported, however, for which intermediate concentrations and applied volumes are most effective[92, 108]. These differences may be related to the form and surface properties of the treated crops. Ford and Reay[92] have studied the biological response of larvae of *Spodoptera littoralis* (Boisd.) to cypermethrin applied in the field at ULV. On cotton the maximum response was obtained at low droplet densities but high concentration of active ingredient per deposit, whereas on lucerne the maximum effectiveness was observed at high droplet density (and low concentration); this suggests that for some crops (e.g. lucerne) extensive cover is more important than the concentration of the active ingredient per unit area, whereas for others (e.g. cotton) the extent of cover is relatively unimportant. These results may be related to the nature of the plant surface (the degree of waxiness or surface roughness) that may modify the availability of the active ingredient to a pest and so influence field efficacy. A low encounter probability between pest and deposit, for example, may result in effective treatment if the proportion of material transferred to the pest on encounter is large. In contrast, if a very small proportion of material is transferred, as might be expected for pickup of oily deposits from waxy surfaces in which the insecticide is dissolved, much higher encounter probabilities would be necessary to ensure that the insect accumulated a toxic dose.

Field studies suggest ways in which the formulation and application of insecticides may be modified to improve performance. At present, the design of application procedures is based largely on empirical observations made during field trials. These inevitably reflect the preliminary ideas about the appropriate set of spray properties available to the investigator at the design stage of the trial, an example of a tail wagging a rather expensive dog. An alternative approach is to determine, prior to application, the pattern and properties of the insecticide deposit on the plant surface that are necessary for maximum effectiveness and to let these properties determine the formulation and spray characteristics. The problem then becomes one of development of methods by which the desired deposit can be achieved. This approach is likely to gain favour as new techniques of analysis lead to a better understanding of the processes underlying the field performance of insecticides. Whether it can eventually replace the strategy based upon empiricism, however, only time will tell.

2.5 Conclusions

The effectiveness of an insecticide used in crop protection will depend on the efficiency of transfer of the active ingredient from the spray nozzle to the plant surface and the efficiency of transfer from that surface to the target insect. Transfer of insecticide to the intermediate plant surface will be subject to the forces of gravity and wind turbulence. The influence of gravity can be modified to assist this process by adjusting the size of the spray droplets and the density of the carrier. Although wind conditions are largely outside the control of the spray operator, crop penetration and subsequent deposition on upward- and downward-facing surfaces can be increased by the use of fan-assisted sprayers. The limitations imposed by gravity and wind restrict the use of very small drops (80 μm diameter and below) which might provide a more effective treatment once retained on the plant. Small drops do not penetrate readily into the crop canopy but are carried away from the crop by upward and lateral drift in prevailing air currents: this is exacerbated where thermal gradients cause upward movement of warm air. The use of vertically displaced or wind-assisted, electrostatically charged, spray particles provides a novel means of reducing this loss by increasing the probability of deposition on the target crop.

Once retained on a plant surface, a deposit will be subject to a number of sources of loss, including penetration into the plant cuticle; evaporative loss of surface deposits to the atmosphere; weathering by the processes of photodegradation, leaching and mechanical detachment due to rainfall; and biotransformation.

Before pickup can occur the target organism must encounter the active ingredient, the probability of which will be reduced by loss processes such as those described above and by avoidance behaviour by the target insect. The probability of encounter will be increased, however, by processes such as diffusion and droplet spread which distribute the active ingredient more evenly over the plant surface: such processes can be critical in determining the ultimate efficacy of a particular treatment and are, therefore, worthy of a more detailed investigation. The insecticide must finally be transferred to the target insect and penetrate to the site of action. Adhesion to the insect cuticle appears to be the major factor determining the transfer of material from plant to insect surface. As long as the force of adhesion between the insect surface and the deposit exceeds that between the deposit and the plant surface, transfer to the insect is favoured.

The effectiveness of any insecticide will therefore depend on the amount that reaches the intermediate surface and on the evenness of distribution on that surface. Unless a variety of reception sites are required on the plant, the use of deposits of uniform size is desirable. However, each deposit should contain sufficient active ingredient to ensure that transfer from the set of deposits likely to be encountered by a target insect will result in death. This can be achieved by adjusting the concentration of active ingredient in the spray, the deposit size and the number per unit surface area. For maximum control, the deposit should be coherent (i.e. free from discontinuities or breaks in cover), a condition approached as the deposit size decreases and the deposit density increases. Much of the available evidence supports this contention although notable exceptions have been reported, particularly for mobile insect pests (e.g. caterpillars). In such cases, a reduction in the deposit size accompanied by an increase in deposit density can dramatically reduce the amount of active ingredient required for a desired level of control by increasing the deposit cover.

Optimum insecticide performance will also depend upon the use of appropriate formulations. These can be designed to maximize both the retention of deposits and the spread of droplets over the plant surface. Droplet spread is particularly important for treatments designed to control immobile stages of the life cycle. For contact action against mobile pests, deposits should be retained in fluid form proud of the epicuticular wax, or as solids or dusts lying loosely upon the surface, whence they may be transferred by adhesion to the insect surface. These considerations suggest that, apart from low-volume use with wettable powders or high-volume ECs or WPs where dry deposits are required, the use of volatile carriers such as water, though inexpensive, probably result in less effective treatments than could be obtained with carefully selected non-volatile, highly viscous and hydrophobic organic solvents. Non-volatile carriers will ensure that droplets retain their size and physical state between atomization and deposition, resulting in evenly distributed sites of insecticide of predetermined cover.

In principle, crop protection problems are best solved by treatments designed for particular applications. Although at present this approach is not entirely feasible, the increased availability of powerful computers should enable the detailed modelling of pesticide–crop interactions to be undertaken so that the properties of a treatment necessary for maximum efficacy can be identified. With appropriate developments in application techniques, it should be feasible to design optimum spray deposition patterns for particular applications and to achieve these under field conditions. Our success in developing this strategy will influence the course of chemical control of phytophagous insect pests over the next few decades. With the advent of insecticides such as the pyrethroids and the avermectins, which possess extremely high levels of activity, a matching performance is required of application and formulation procedures. Application technology should therefore have a major impact on the future development of pest control practices.

2.6 Acknowledgements

We are grateful to a number of colleagues for their valuable assistance during the preparation of this chapter. Cliff Hart (ICI Plant Protection, Jealotts Hill) and Gary Crease (Department of Biological Sciences, Portsmouth Polytechnic) provided scanning electron micrographs; Nigel Scopes, Ian Wyatt and their colleagues at GCRI Rustington, Littlehampton, allowed us to preview the results of their studies (now published) on the biocidal activities of ULV deposits; Alister Hill (Shell Research, Sittingbourne) and Dick Reay (Department of Biological Sciences, Portsmouth Polytechnic) read early drafts of the manuscript and offered constructive comments; and Janet Sugden spent many painstaking hours word-processing in order to revise and update the text.

2.7 References

1. G.S. Hartley and I.J. Graham-Bryce, *Physical Principles of Pesticide Behaviour*, Vol 1 and 2, Academic Press, London (1980).
2. C.G.L. Furmidge, 'General principles governing the behaviour of granular formulations', *Pestic. Sci.*, **3**, 745–51 (1972).
3. R.A. Coffee, 'Electrodynamic crop spraying', *Outlook on Agriculture*, **10**, 350–6 (1981).
4. A.J. Arnold and B.J. Pye, 'Electrostatic spraying of crops with the Ape 80', *Proceedings 1981 British Crop Protection Conference on Pests and Diseases* (1981).
5. D.R. Johnstone and K.A. Huntington 'Deposition and drift of ULV and VLV insecticide sprays applied to cotton by hand applicator in Northern Nigeria', *Pestic. Sci.*, **8**, 101–9 (1977).
6. J.J. Spillmann, 'Spray impaction, retention and adhesion: An introduction to basic characteristics', *Pestic. Sci.*, **15**, 97–106 (1984).
7. G.A. Matthews, *Pesticide Application Methods*, Longman (1979).
8. W. Maas, *ULV Applications and Formulation Techniques*, N.V. Philips-Duphar (1971).
9. W. van Valkenburg *Pesticide Formulations*, Marcel Dekker, New York (1973).
10. G.J. Crease, M.G. Ford and D.W. Salt, 'Studies of the relationship between the properties of carrier solvents and biological efficacy of ULV applied drops of the insecticide cypermethrin', British Crop Protection Monogram No. 28, Symposium on Application and Biology (1985), pp. 251–8.
11. A.W.A. Brown, *Insect Control by Chemicals*, Wiley, New York (1951).
12. P.J. Holloway, 'Surface factors affecting the wetting of leaves', *Pestic. Sci.*, **1**, 156–63 (1970).

13. C.E. Jeffree, E.A. Baker and P.J. Holloway 'Origins of the fine structure of plant epicuticular waxes', in *Microbiology of Aerial Plant Surfaces*, ed. by C.H. Dickenson and T.F. Preece, Academic Press, London (1976), pp. 119–58.

14. N.K. Adam and J. Jessop 'Angles of contact and polarity of solid surfaces', *J. Chem. Soc.*, **127**, 1863–8 (1925).

15. R.N. Wensel, 'Resistance of solid surfaces to wetting by water', *Ind. Eng. Chem.*, **28**, 988–94 (1936).

16. D.R. Kreger, 'An X-ray study of waxy coatings from plants', *Recl. Trav. Bot. Neerl.*, **41**, 603–736 (1948).

17. S.H. Piper, A.C. Chibnall, S.J. Hopkins, A. Pollards, J.A.B. Smith and E.F. Williams, 'Synthesis and crystal spacings of certain long-chain paraffins, ketones and secondary alcohols', *Biochem. J.*, **25**, 2072–94 (1933).

18. D.R. Kreger and C. Schamhart, 'On the long crystal-spacings in wax esters and their value in micro-analysis of plant cuticle waxes', *Biochem. Biophys. Acta*, **19**, 22–44 (1956).

19. E.A. Baker, G.M. Hunt and P.J.G. Stevens, 'Studies of plant cuticle and spray droplet interactions: A fresh approach', *Pestic. Sci.*, **14**, 645–58 (1983).

20. D.W. Salt and M.G. Ford, 'The kinetics of insectide action. Part III: The use of stochastic modelling to investigate the pick-up of insecticides from ULV-treated surfaces by larvae of *Spodoptera littoralis* Boisd., *Pestic. Sci.*, **15**, 382–410 (1984).

21. A.B.D. Cassie and S. Baxter, 'Wettability of porous surfaces', *Trans. Faraday Soc.*, **40**, 546–51 (1944).

22. S.B. Challen, 'The contribution of surface characters to the wettability of leaves', *J. Pharm. Pharmacol.* **12**, 307–11 (1960).

23. D.M. Hall, 'A study of the surface wax deposits on apple fruit', *Aust. J. Biol. Sci.*, **19**, 1017–25 (1966).

24. A.M.S. Fernandes, E.A. Baker and J.T. Martin, 'Studies on plant cuticle. VI. The isolation and fractionation of cuticular waxes', *Ann. Appl. Biol.*, **53**, 43–58 (1964).

25. D.M. Hall, A.I. Matus, J.A. Lamberton and H.N. Barber, 'Infra-specific variation in wax on leaf surfaces', *Aust. J. Biol. Sci.*, **18**, 323–32 (1965).

26. W. Gukel and G. Synnatshke, 'Techniques for measuring the wetting of leaf surfaces', *Pestic. Sci.*, **6**, 595–603 (1975).

27. C.A. Hart, 'Use of the scanning electron microscope and cathodoluminescence in studying the application of pesticides to plants', *Pestic. Sci.*, **10**, 341–57 (1979).

28. R.F. Norris and M.J. Bukovac, 'Influence of cuticular waxes on penetration of pear leaf cuticle by l-naphthaleneacetic acid', *Pestic. Sci.*, **3**, 705–708 (1972).

29. D.G. Davies, J.S. Mullins, G.E. Stolsenberg and G.D. Booth, 'Permeation of organic molecules of widely differing solubilities and of water through isolated cuticles of orange leaves', *Pestic. Sci.*, **10**, 19–31 (1979).

30. A.G.L. Wilson, J.M. Desmarchelier and K. Malafant, 'Persistance on cotton foliage of insecticide residues toxic to *Heliothis* larvae', *Pestic. Sci.*, **14**, 623–33 (1983).

31. F.T. Phillips and E.M. Gillham 'Persistence to rainwashing of DDT wettable powders', *Pestic. Sci.*, **2**, 97–100 (1971).

32. N.H. Anderson and J. Girling 'The uptake of surfactants into wheat', *Pestic. Sci.*, **14**, 399–404 (1983).

33. F.T. Phillips, P. Etheridge, V.S. Kavadia, G.R. Sethi and P.E. Sparrow, 'Translocation of ^{14}C-Dieldrin from small droplets on cotton leaves', *Ann. Appl. Biol.*, **89**, 51–9 (1978).

34. G.G. Briggs, R.H. Bromilow and A.A. Evans, 'Relationships between lipophilicity and root uptake and translocation of non-ionised chemicals by barley', *Pestic. Sci.*, **13**, 495–504 (1982).

35. F.T. Phillips, 'Persistence of organochlorine insecticides on different substrates under different environmental conditions. I. The rates of loss of dieldrin and aldrin by volatilisation from glass surfaces', *Pestic. Sci.*, **2**, 255–66 (1971).

36. F.T. Phillips, 'Some aspects of volatilisation of organochlorine insecticides', *Chem. Ind.*, 193–7 (1974).

37. F.E. Pick, L.P. van Dyke and P.R. de Beer, 'The effect of simulated rain on deposits of some cotton pesticides', *Pestic. Sci.*, **15**, 616–23 (1984).

38. S.J. Nemec and P.J. Adkisson 'Effects of simulated rain and dew on the toxicity of certain ultra-low-volume insecticidal spray formulations', *J. Econ. Entomol.*, **62**(1), 71–3 (1969).

39. G.G. Briggs, 'Degradation in soils', Proceedings of BCPC Symposium entitled *Persistence of Insecticides and Herbicides* (1976).

40. T.F. Preece and C.H. Dickinson *Ecology of Leaf Surface Microorganisms*, Academic Press (1971).
41. C.H. Dickinson and T.F. Preece, *Microbiology of Aerial Plant Surfaces*, Academic Press (1976).
42. J.I. Williams and G.J.F. Pugh, 'Restistance of *Chrysosporium pannorum* to an organomercury fungicide', *Trans. Br. Mycol. Soc.*, **64**, 255–63 (1975).
43. A.J. Kathubutheen and G.J.F. Pugh, 'Effects of temperature and fungicides on *Chrysosporium pannorum* (Link Hughes)', *Trans. Br. Mycol. Soc.*, **45**, 65–79 (1979).
44. A.J. Kathubutheen, 'The effects of fungicides on soil and leaf fungi', PhD thesis, University of Aston (1977).
45. D.R. Johnstone, 'Spreading and retention of agricultural sprays on foliage', in *Pesticide Formulations*, ed. by W. van Valkenburg, Marcel Dekker, New York (1973).
46. C.T. Lewis, J.C. Hughes, 'Studies concerning the uptake of contact insecticides. II. The contamination of flies exposed to particulate deposits', *Bul. Ent. Res.*, **48**, 755–68 (1957).
47. M. Gratwick, 'The uptake of DDT and other lipophilic particles by blowflies walking over deposits', *Bul. Ent. Res.*, **48**, 733–40 (1957).
48. G. Armstrong, F.R. Bradbury and H. Standen, 'The penetration of the insect cuticle by isomers of benzene hexachloride', *Ann. Appl. Biol.*, **38**, 555–66 (1951).
49. G. Armstrong, F.R. Bradbury and H.G. Britton, 'The Penetration of the insect cuticle by DDT and related compounds', *Ann. Appl. Biol.*, **39**, 548–56 (1952).
50. P.T. Brey, H. Ohayon, M. Lesourd, H. Castex, J. Roucache and J.P. Latge, 'Ultrastructure and chemical composition of the outer layers of the cuticle of the pea aphid *Acyrthosiphon pisum* (Harris)', *Comp. Biochem. Physiol.*, **82A**(2), 401–11 (1958).
51. N.F. Hadley, 'Cuticular lipids of terrestrial plants and arthropods: A comparison of their structure, composition, and water-proofing function', *Biol. Rev.*, **56**, 23–47 (1981).
52. G.J. Blomquist and L.L. Jackson 'Chemistry and biochemistry of insect waxes', *Prog. Lipid Res.*, **17**, 319–45 (1979).
53. E.L. Cussler, *Diffusion — Mass Transfer in Fluid Systems*, Cambridge University Press (1984).
54. L.P. Schouest, N. Umetsu and T.A. Miller, 'Solvent-modified deposition of insecticides on house fly (Diptera: *Muscidae*) cuticle', *J. Econ. Entomol.*, **76**, 973–82 (1983).
55. S.L. Gowers and G.N.J. Le Patourel, 'Toxicity of deposits of an amorphous silica dust on different surfaces and their pick-up by *Sitophilus granarius* (L)', *J. Stored Prod. Res.*, **20**, 25–9 (1984).
56. A.B. Hadaway and F. Barlow, 'The influence of environmental conditions on the contact toxicity of some insecticide deposits to adult mosquitos, *Anopheles stephensi* (List)', *Bul. Ent. Res.*, **54**, 329–43 (1963).
57. J.E. Beesley and P.R. Chadwick, Abstract from a talk entitled 'A scanning electron microscope study of surface deposits of insecticides' presented at a symposium on *Insecticidal Deposits: Pickup and Biological Activity* organized jointly by the Association of applied Biologists and the Society of Chemical Industry (1982).
58. P.R. Chadwick and S.W. Carter Abstract from a talk entitled 'Aspects of residual deposit activity' presented at a symposium on *Insecticidal Deposits: Pickup and Biological Activity* organized jointly by the Association of Applied Biologists and the Society of Chemical Industry (1982).
59. P.R. Chadwick, 'Surfaces and other factors modifying the effectiveness of pyrethroids against insects in public health', *Pestic. Sci.*, **16**, 383–91 (1985).
60. P.E. Burt and J. Ward, 'The persistence and fate of DDT on foliage. I — The influence of plant wax on the toxicity and persistence of deposits of DDT crystals', *Bul. Ent. Res.*, **46**, 39–56 (1955).
61. P. Gerolt, 'Influence of relative humidity on the uptake of insecticides from residual films', *Nature, London*, **197**, 721 (1963).
62. V.B. Wigglesworth, *The Principles of Insect Physiology*, 7th ed., Chapman and Hall (1972).
63. D.R. Johnstone, 'Formulations and atomisation', in *Proceedings of 4th International Agricultural Aviation Congress* (Kingston, 1969), p. 225.
64. D.R. Johnstone, 'Development of a technique for the laboratory application of deposits of low and ultra-low volume character', Tropical Pesticide Research Unit Rept. for 1969/70, pp. 11–18 (1971).
65. A.B. Hadaway and F. Barlow 'Studies of aqueous suspensions of insecticides', *Bul. Ent. Res.*, **41**, 603–22.

66. M.G. Ford, R.C. Reay and W.S. Watts, 'Laboratory evaluation of the activity of synthetic pyrethroids at ULV against the cotton leaf worm *Spodoptera littoralis* Boisd.', in *Crop Protection Agents — Their Biology Evaluation* ed. by N.R. McFarlane, Academic Press (1977).

67. R.C. Reay and M.G. Ford, 'Toxicity of pyrethroids to larvae of the Egyptian cotton leaf worm, *Spodoptera littoralis* (Boisd.)., II. Factors determining the effectiveness of permethrin at ULV', *Pestic. Sci.*, **8**, 243–53 (1977).

68. N.W. Wooten and K.F. Sawyer, 'The pick-up of spray droplets by flying locusts', *Bul. Ent. Res.*, **45**, 177–97 (1954).

69. N. Thompson, 'Diffusion and uptake of chemical vapour volatilising from a sprayed target area', *Pestic. Sci.*, **14**, 33–9 (1983).

70. I.J. Graham-Bryce, J.H. Stevenson and P. Etheridge, 'Factors affecting the performance of granular insecticides applied to field beans', *Pestic. Sci.*, **3**, 781–97 (1972).

71. M. Abdalla, 'A biological study of the spread of pesticides from small droplets', PhD thesis, University of London (1984).

72. C.R. Worthing, *The Pesticide Mannual, A World Compendium*, 6th ed, The British Crop Protection Council (1979).

73. D.D. Evans, Abstract from a talk entitled 'Routes of insecticide toxicity related to their field efficacy' presented at a symposium on *Insecticidal Deposits: Pickup and Biological Activity* organized jointly by the Association of Applied Biologists and the Society of Chemical Industry (1982).

74. R.N. Price, Shell Research Ltd. Sittingbourne, Kent, personal communication.

75. P.J.G. Stevens, Abstract from a talk entitled 'Physicochemical parameters affecting uptake of foliar applied ^{14}C-chemicals from spray droplets' presented at a symposium on *The Delivery of Pesticides and Their Behaviour on the Target* organized by the Society of Chemical Industry (1983).

76. R.A. Dybas and A. St Green, 'Avermectins: Their chemistry and pesticidal activity', in Proceedings of the 1984 British Crop Protection Conference on *Pests and Diseases*, Vol. 3 (1984), pp. 947–54.

77. C.N.E. Ruscoe, 'The new NRDC pyrethroids as agricultural insecticides', *Pestic. Sci.*, **8**, 236–42 (1977).

78. K-H. Tan, 'Antifeeding effect of cypermethrin and permethrin at sublethal levels against *Pieris brassicae* larva, *Pestic. Sci.*, **12**, 619–26 (1981).

79. J.A. Schemanchuk, 'Repellent action of permethrin, cypermethrin and resmethrin against black flies (*Simulium* spp.) attacking cattle', *Pestic. Sci.*, **12**, 412–16 (1981).

80. K.F. Armstrong and A.B. Bonner, 'Investigation of a permethrin-induced antifeedant effect in *Drosophila melanogaster*: An ethical approach', *Pestic. Sci.*, **16**, 641–50 (1985).

81. A. Antunes de Almeida, 'Reactions of larvae of *Plodia interpunctella* (HB) (LEP, *Pyralidae*) to insecticidal droplets', *Bul. Ent. Res.*, **58**, 221–6 (1967).

82. Roussel Uclaf, Deltamethrin Monograph (1982).

83. S.W. Shires, A. Murray, P. Debray and J. Le Blanc, 'The effects of a new pyrethroid insecticide WL–85871 on foraging honey bees (*Apis mellifera* L.) *Pestic. Sci.*, **15**, 491–9 (1984).

84. A.D. Rice, R.W. Gibson and M.F. Stribley, 'Alarm pheromone secretion by insecticide-suscepti-ble and -resistant *Myzus persicae* treated with demeton-S-methyl; aphid dispersal and transfer of plant viruses', *Ann. Appl. Biol.*, **103**, 375–81 (1983).

85. S.G. Polles and S. Bradleigh Vinson, 'Effect of droplet size on persistence of ULV malathion and comparison of toxicity of ULV and EC malathion to tobacco budworm larvae', *J. Econ. Entomol.*, **62**, 89–94 (1969).

86. S. Miall and I.F. Turner, Abstract from a talk entitled 'The irritancy of insecticidal deposits to insects: Residual pyrethroids vs. *Blattella germanica*' presented at a symposium on *Insecticidal Deposits: Pickup and Biological Activity* organized jointly by the Association of Applied Biologists and the Society of Chemical Industry (1982).

87. K.B. Wildey, Abstract from a talk entitled 'A formulation repellancy of stored product insecticides in a resistant and a susceptible strain of *Tribolium castaneum*' presented at a symposium on *Insecticidal Deposits: Pickup and Biological Activity* organised jointly by the Association of Applied Biologists and the Society of Chemical Industry (1982).

88. G. Hoyle, 'Understanding the cellular basis of insect behaviour', in Proceedings of a Society of Chemical Industry symposium on *Insect Neurobiology and Pesticide Action* (1979), pp. 399–406.

89. W. Ebeling, R.E. Wagner and D.A. Reierson, 'Influence of repellency on the efficacy of blatticides. I. Learned modification of behaviour of the German cockroach', *J. Econ. Entomol.*, **59**, 1374–88.

90. W. Ebeling, D.A. Reierson and R.E. Wagner, 'The influence of repellency on the efficacy of blatticides. III. Field experiments with German cockroaches with notes on three other species' *J. Econ. Entomol.*, **61**, 751–61 (1968).

91. M.K. Rust and D.A. Reierson, 'Comparsion of the laboratory and field efficacy of insecticides used for German cockroach control', *J. Econ. Entomol.*, **71**, 704–8 (1978).

92. M.G. Ford and R.C. Reay, unpublished results of field trials undertaken in Morocco (1977).

93. N.E.A. Scopes, 'Some factors affecting the efficiency of small pesticide droplets', in Proceedings of the 1981 British Crop Protection Conference on *Pests and Diseases* (1981).

94. D.C. Munthali, 'Biological efficiency of small pesticide droplets', PhD thesis, University of London (1981).

95. D.C. Munthali and N.E.A Scopes, 'A technique for studying the biological efficiency of small droplets of pesticide solutions and a consideration of the implications', *Pestic. Sci.*, **13**, 60–2 (1982).

96. E.J. Bals, 'The principles of and new developments in ULV spraying: some reflections', in Proceedings of the 1982 British Crop Protection Conference on *Weeds* (1982).

97. A. Palmer, I.J. Wyatt and N.E.A Scopes, 'The toxicity of ULV permethrin to Glasshouse whitefly', ICPP Research Rep. 3A–R16, Glasshouse Research Institute (1983).

98. I.J. Wyatt, M.R. Abdalla, P.T. Atkey and A. Palmer, 'Activity of discrete permethrin droplets against whitefly scales', British Crop Protection Conference on *Pests and Diseases*, Vol. 13 (1984), pp. 1045–8.

99. I.J. Wyatt, M.R. Abdalla, A. Palmer and D.C. Munthali, 'Localized activity of ULV pesticide droplets against sedentary pests', Syposium on *Application and Biology*, British Crop Protection Monogram No. 28 (1985), pp. 259–64.

100. S. Uk and R.J. Courshee, 'Distribution and likely effectiveness of spray deposits within a cotton canopy from fine ultralow-volume spray applied by aircraft', *Pestic. Sci.*, **13**, 529–536 (1982).

101. A.J. Arnold, G.R. Cayley, Y. Dunne, D. Etheridge, D.C. Griffiths, F.T. Phillips, B.J. Pye, G.C. Scott and P.R. Vojvodic, 'Biological effectiveness of electrostatically charged rotary atomisers. I. Trials on field beans and barley, 1981', *Ann. Appl. Biol.*, **105**, 353–9 (1984).

102. A.J. Arnold, G.R. Cayley, Y. Dunne, P. Etheridge, D.C. Griffiths, J.F. Jenkyn, F.T. Phillips, B.J. Pye, G.C. Scott and C.M. Woodcock, 'Biological effectiveness of electrostatically charged rotary atomisers. II. Trials with cereals, 1982', *Ann. Appl. Biol.*, **105**, 361–7 (1982).

103. A.J. Arnold, G.R. Cayley, Y. Dunne, P. Etheridge, A.R. Greenway, D.C. Griffiths, F.T. Phillips, B.J. Pye, C.J. Rawlinson and G.C. Scott, 'Biological effectiveness of electrostatically charged rotary atomisers. III. Trials on arable crops other than cereals, 1982', *Ann. Appl. Biol.*, **105**, 369–77 (1984).

104. G.R. Cayley, P. Etheridge, D.C. Griffiths, F.T. Phillips, B.J. Pye and G.C. Scott, 'A review of the performance of electrostatically charged rotary atomisers on different crops', *Ann. Appl. Biol.*, **105**, 379–86 (1984).

105. D.C. Griffiths, G.R. Cayley, P. Etheridge, R.E. Goodchild, P.J. Hulme, R.J. Lewthwaite, B.J. Pye, G.C. Scott and J.H. Stevenson, 'Application of insecticides, fungicides and herbicides to cereals with charged rotary atomisers', 1984 British Crop Protection Conference on *Pests and Diseases* (1984), pp. 1021–6.

106. G.R. Cayley, P.E. Etheridge, R.E. Goodhild, D.C. Griffiths, P.J. Hulme, R.J. Lewthwaite, B.J. Pye and G.C. Scott, 'Review of the relationship between chemical deposits achieved with electro-statically charged rotary atomisers and their biological effects', Symposium on *Application and Biology*, PCPC Monogram No. 28 (1985), pp. 87–96.

107. F.R. Hall and D.L. Reichard, 'Effects of spray droplet size, dosage and solution per ha on Rates of mortality of twospotted spider mite', *J. Econ. Entomol.*, **71**, 279–82 (1978).

108. D.R. Johnstone, 'Insecticide concentration for ultra-low volume crop spray application', *Pestic. Sci.*, **4**, 77–82 (1973).

Index